Equations of State
A Graphical Comparison

by D. James Benton

Copyright © 2022 by D. James Benton, all rights reserved.

Preface

This is a compilation of 160 graphs showing how 16 different equations of state compare in their prediction of 10 different thermodynamic properties. Only minimal discussion of the details is provided herein, as this is readily available in many Web articles and texts, including my, *Thermodynamic & Transport Properties of Fluids*. What we see is that there are strengths and weaknesses for many of these equations, especially the early and simple ones. The greatest refinement and effort has been given to the 2020 Steam Formulation, as is evident in the figures.

All of the examples contained in this book,
(as well as a lot of free programs) are available at…
https://www.dudleybenton.altervista.org/software/index.html

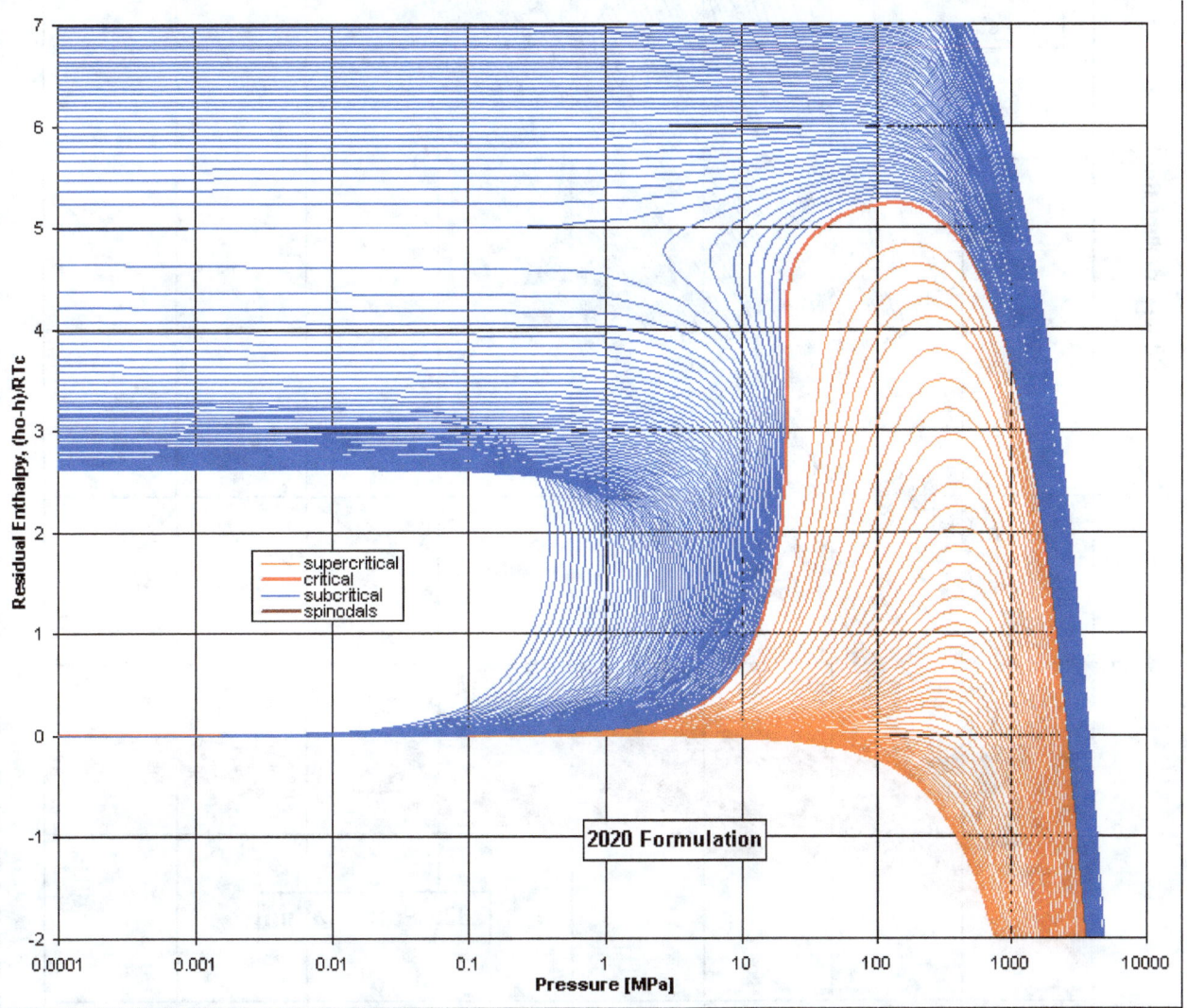

Figure 1. Residual Enthalpy vs. Pressure Based on Steam 2020 Formulation

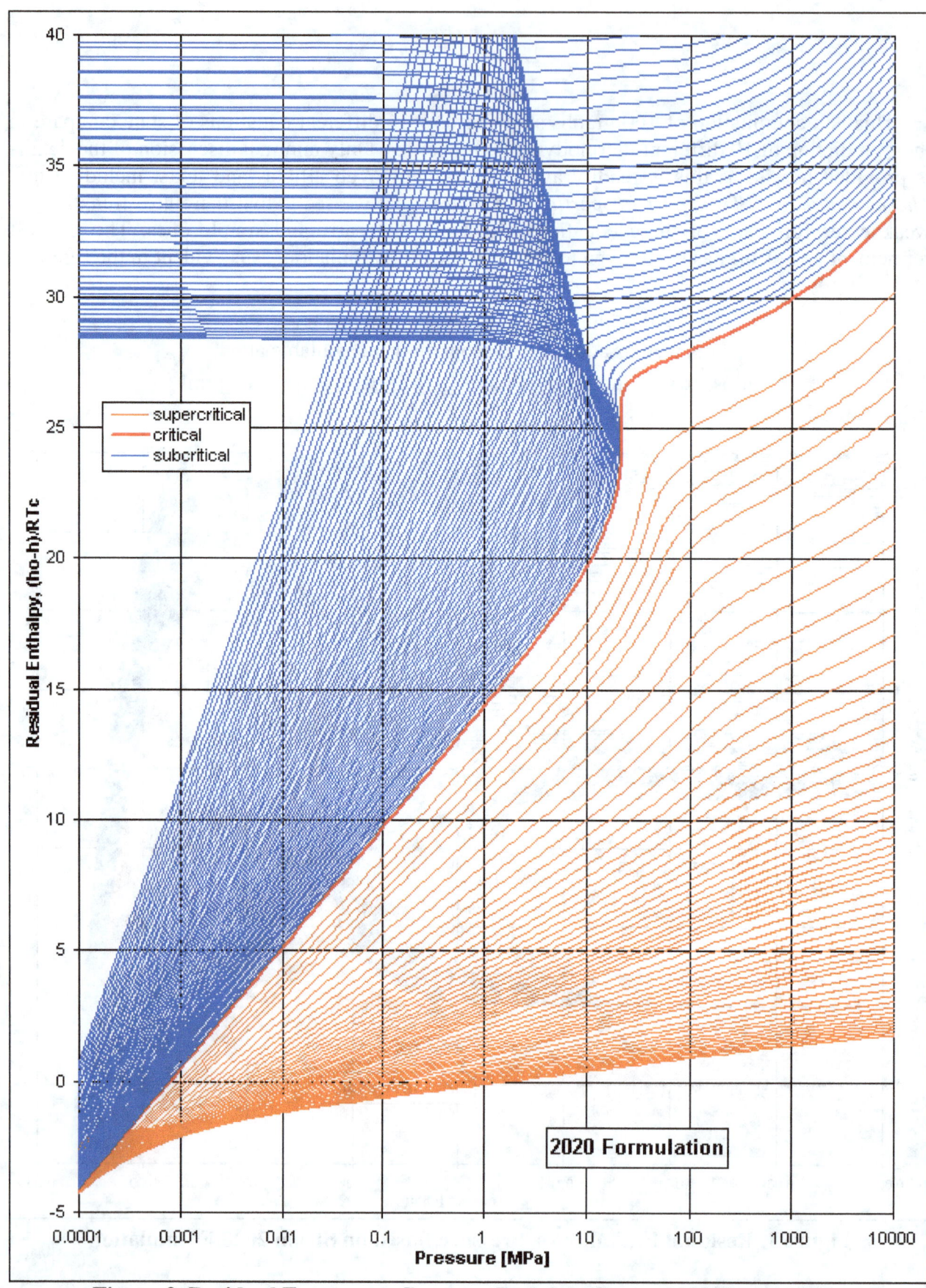

Figure 2. Residual Entropy vs. Pressure Based on Steam 2020 Formulation

Table of Contents

	page
Preface	i
Chapter 1. Introduction	1
Chapter 2. Pressure vs. Volume	5
Chapter 3. Temperature vs. Pressure	25
Chapter 4. Compressibility	41
Chapter 5. Fugacity	57
Chapter 6. Residual Enthalpy	75
Chapter 7. Residual Entropy	93
Chapter 8. Pressure vs. Enthalpy	111
Chapter 9. Temperature vs. Entropy	127
Chapter 10. Mollier Chart	143
Chapter 11. ZT_R vs. $1/V_R$	159

List of Figures

	page
Figure 1. Residual Enthalpy vs. Pressure Based on Steam 2020 Formulation	i
Figure 2. Residual Entropy vs. Pressure Based on Steam 2020 Formulation	ii
Figure 3. van der Waals Equation of State	2
Figure 4. Maxwell's Criterion of Equal Areas	3
Figure 5. Pr vs. Vr Based on Nelson-Obert	8
Figure 6. Pr vs. Vr Based on van der Waals	9
Figure 7. Pr vs. Vr Based on Boltzmann	10
Figure 8. Pr vs. Vr Based on Berthelot	11
Figure 9. Pr vs. Vr Based on Deterici	12
Figure 10. Pr vs. Vr Based on Clausius	13
Figure 11. Pr vs. Vr Based on Abbott's Modification	14
Figure 12. Pr vs. Vr Based on Redlich-Kwong	15
Figure 13. Pr vs. Vr Based on Soave's Modification	16
Figure 14. Pr vs. Vr Based on Fuller's Modification	17
Figure 15. Pr vs. Vr Based on Peng-Robinson	18
Figure 16. Pr vs. Vr Based on Author's Modification	19
Figure 17. Pr vs. Vr Based on Keenan, Keyes, Hill, and Moore	20
Figure 18. Pr vs. Vr Based on Haar, Gallagher, and Kell	21
Figure 19. Pr vs. Vr Based on Wagner and Pruß	22
Figure 20. Pr vs. Vr Based on Steam 2020 Formulation	23
Figure 21. T vs. V Based on Nelson-Obert	25
Figure 22. Tr vs. Vr Based on van der Waals	26
Figure 23. Tr vs. Vr Based on Boltzmann	27
Figure 24. Tr vs. Vr Based on Berthelot	28
Figure 25. Tr vs. Vr Based on Dieterici	29
Figure 26. Tr vs. Vr Based on Clausius	30
Figure 27. Tr vs. Vr Based on Abbott's Modification	31
Figure 28. Tr vs. Vr Based on Redlich-Kwong	32
Figure 29. Tr vs. Vr Based on Soave's Modification	33
Figure 30. Tr vs. Vr Based on Fuller's Modification	34
Figure 31. Tr vs. Vr Based on Peng-Robinson	35
Figure 32. Tr vs. Vr Based on Author's Modification	36

Figure 33. Tr vs. Vr Based on Keenan, Keyes, Hill, and Moore 37
Figure 34. Tr vs. Vr Based on Haar, Gallagher, and Kell 38
Figure 35. Tr vs. Vr Based on Wagner and Pruß 39
Figure 36. Tr vs. Vr Based on Steam 2020 Formulation 40
Figure 37. Compressibility Based on Nelson-Obert 41
Figure 38. Compressibility Based on van der Waals 42
Figure 39. Compressibility Based on Boltzmann 43
Figure 40. Compressibility Based on Berthelot 44
Figure 41. Compressibility Based on Dieterici 45
Figure 42. Compressibility Based on Clausius 46
Figure 43. Compressibility Based on Abbott's Modification 47
Figure 44. Compressibility Based on Redlich-Kwong 48
Figure 45. Compressibility Based on Soave's Modification 49
Figure 46. Compressibility Based on Fuller's Modification 50
Figure 47. Compressibility Based on Peng-Robinson 51
Figure 48. Compressibility Based on Author's Modification 52
Figure 49. Compressibility Based on Keenan, Keyes, Hill, and Moore 53
Figure 50. Compressibility Based on Haar, Gallagher, and Kell 54
Figure 51. Compressibility Based on Wagner and Pruß 55
Figure 52. Compressibility Based on Steam 2020 Formulation 56
Figure 53. Fugacity Coefficient Based on the Nelson-Obert Generalized Data 58
Figure 54. Fugacity Coefficient Based on van der Waals 59
Figure 55. Fugacity Coefficient Based on Boltzmann 60
Figure 56. Fugacity Coefficient Based on Berthelot 61
Figure 57. Fugacity Coefficient Based on Dieterici 62
Figure 58. Fugacity Coefficient Based on Clausius 63
Figure 59. Fugacity Coefficient Based on Abbott's Modification 64
Figure 60. Fugacity Coefficient Based on Redlich-Kwong 65
Figure 61. Fugacity Coefficient Based on Soave's Modification 66
Figure 62. Fugacity Coefficient Based on Fuller's Modification 67
Figure 63. Fugacity Coefficient Based on Peng-Robinson 68
Figure 64. Fugacity Coefficient Based on Author's Modification 69
Figure 65. Fugacity Coefficient Based on Keenan, Keyes, Hill, and Moore 70
Figure 66. Fugacity Coefficient Based on Haar, Gallagher, and Kell 71
Figure 67. Fugacity Coefficient Based on Wagner and Pruß 72
Figure 68. Fugacity Coefficient Based on Steam 2020 Formulation 73
Figure 69. Residual Enthalpy Based on Nelson-Obert 76
Figure 70. Residual Enthalpy Based on van der Waals 77
Figure 71. Residual Enthalpy Based on Boltzmann 78
Figure 72. Residual Enthalpy Based on Berthelot 79
Figure 73. Residual Enthalpy Based on Dieterici 80
Figure 74. Residual Enthalpy Based on Clausius 81
Figure 75. Residual Enthalpy Based on Abbott's Modification 82
Figure 76. Residual Enthalpy Based on Redlich-Kwong 83
Figure 77. Residual Enthalpy Based on Soave's Modification 84
Figure 78. Residual Enthalpy Based on Fuller's Modification 85
Figure 79. Residual Enthalpy Based on Peng-Robinson 86
Figure 80. Residual Enthalpy Based on Author's Modification 87
Figure 81. Residual Enthalpy Based on Keenan, Keyes, Hill, and Moore 88

Figure 82. Residual Enthalpy Based on Haar, Gallagher, and Kell 89
Figure 83. Residual Enthalpy Based on Wagner and Pruß 90
Figure 84. Residual Enthalpy Based on Steam 2020 Formulation 91
Figure 85. Residual Entropy Based on Nelson-Obert 94
Figure 86. Residual Entropy Based on van der Waals 95
Figure 87. Residual Entropy Based on Boltzmann 96
Figure 88. Residual Entropy Based on Berthelot 97
Figure 89. Residual Entropy Based on Dieterici 98
Figure 90. Residual Entropy Based on Clausius 99
Figure 91. Residual Entropy Based on Abbott's Modification 100
Figure 92. Residual Entropy Based on Redlich-Kwong 101
Figure 93. Residual Entropy Based on Soave's Modification 102
Figure 94. Residual Entropy Based on Fuller's Modification 103
Figure 95. Residual Entropy Based on Peng-Robinson 104
Figure 96. Residual Entropy Based on Author's Modification 105
Figure 97. Residual Entropy Based on Keenan, Keyes, Hill, and Moore 106
Figure 98. Residual Entropy Based on Haar, Gallagher, and Kell 107
Figure 99. Residual Entropy Based on Wagner and Pruß 108
Figure 100. Residual Entropy Based on Steam 2020 Formulation 109
Figure 101. Pressure vs. Enthalpy Based on Nelson-Obert 111
Figure 102. Pressure vs. Enthalpy Based on van der Waals 112
Figure 103. Pressure vs. Enthalpy Based on Boltzmann 113
Figure 104. Pressure vs. Enthalpy Based on Berthelot 114
Figure 105. Pressure vs. Enthalpy Based on Dieterici 115
Figure 106. Pressure vs. Enthalpy Based on Clausius 116
Figure 107. Pressure vs. Enthalpy Based on Abbott's Modification 117
Figure 108. Pressure vs. Enthalpy Based on Redlich-Kwong 118
Figure 109. Pressure vs. Enthalpy Based on Soave's Modification 119
Figure 110. Pressure vs. Enthalpy Based on Fuller's Modification 120
Figure 111. Pressure vs. Enthalpy Based on Peng-Robinson 121
Figure 112. Pressure vs. Enthalpy Based on Author's Modification 122
Figure 113. Pressure vs. Enthalpy Based on Keenan, Keyes, Hill, and Moore 123
Figure 114. Pressure vs. Enthalpy Based on Haar, Gallagher, and Kell 124
Figure 115. Pressure vs. Enthalpy Based on Wagner and Pruß 125
Figure 116. Pressure vs. Enthalpy Based on Steam 2020 Formulation 126
Figure 117. Temperature vs. Entropy Based on Nelson-Obert 127
Figure 118. Temperature vs. Entropy Based on van der Waals 128
Figure 119. Temperature vs. Entropy Based on Boltzmann 129
Figure 120. Temperature vs. Entropy Based on Berthelot 130
Figure 121. Temperature vs. Entropy Based on Dieterici 131
Figure 122. Temperature vs. Entropy Based on Clausius 132
Figure 123. Temperature vs. Entropy Based on Abbott's Modification 133
Figure 124. Temperature vs. Entropy Based on Redlich-Kwong 134
Figure 125. Temperature vs. Entropy Based on Soave's Modification 135
Figure 126. Temperature vs. Entropy Based on Fuller's Modification 136
Figure 127. Temperature vs. Entropy Based on Peng-Robinson 137
Figure 128. Temperature vs. Entropy Based on Author's Modification 138
Figure 129. Temperature vs. Entropy Based on Keenan, Keyes, Hill, and Moore 139
Figure 130. Temperature vs. Entropy Based on Haar, Gallagher, and Kell 140

Figure 131. Temperature vs. Entropy Based on Wagner and Pruß 141
Figure 132. Temperature vs. Entropy Based on Steam 2020 Formulation.......................................142
Figure 133. Mollier Chart Based on Nelson-Obert...143
Figure 134. Mollier Chart Based on van der Waals..144
Figure 135. Mollier Chart Based on Boltzmann..145
Figure 136. Mollier Chart Based on Berthelot...146
Figure 137. Mollier Chart Based on Dieterici 147
Figure 138. Mollier Chart Based on Clausius 148
Figure 139. Mollier Chart Based on Abbott's Modification 149
Figure 140. Mollier Chart Based on Redlich-Kwong..150
Figure 141. Mollier Chart Based on Soave's Modification...151
Figure 142. Mollier Chart Based on Fuller's Modification 152
Figure 143. Mollier Chart Based on Peng-Robinson...153
Figure 144. Mollier Chart Based on Author's Modification..154
Figure 145. Mollier Chart Based on Keenan, Keyes, Hill, and Moore...155
Figure 146. Mollier Chart Based on Haar, Gallagher, and Kell 156
Figure 147. Mollier Chart Based on Wagner and Pruß 157
Figure 148. Mollier Chart Based on Steam 2020 Formulation..158
Figure 149. ZTr vs. 1/Vr Based on Nelson-Obert 159
Figure 150. ZTr vs. 1/Vr Based on van der Waals 160
Figure 151. ZTr vs. 1/Vr Based on Boltzmann..161
Figure 152. ZTr vs. 1/Vr Based on Berthelot 162
Figure 153. ZTr vs. 1/Vr Based on Dieterici 163
Figure 154. ZTr vs. 1/Vr Based on Clausius 164
Figure 155. ZTr vs. 1/Vr Based on Abbott's Modification 165
Figure 156. ZTr vs. 1/Vr Based on Redlich-Kwong..166
Figure 157. ZTr vs. 1/Vr Based on Soave's Modification...167
Figure 158. ZTr vs. 1/Vr Based on Fuller's Modification 168
Figure 159. ZTr vs. 1/Vr Based on Peng-Robinson...169
Figure 160. ZTr vs. 1/Vr Based on Author's Modification...170
Figure 161. ZTr vs. 1/Vr Based on Keenan, Keyes, Hill, and Moore 171
Figure 162. ZTr vs. 1/Vr Based on Haar, Gallagher, and Kell 172
Figure 163. ZTr vs. 1/Vr Based on Wagner and Pruß 173
Figure 164. ZTr vs. 1/Vr Based on Steam 2020 Formulation 174

Chapter 1. Introduction

An Equation of State (EOS) in the thermodynamic sense is a mathematical relationship between the pressure, density, and temperature of a fluid (gas or liquid). The reciprocal of density is specific volume, which is often used instead of density in these formulas. The first formally published EOS is attributed to van der Waals in 1873[1].

$$P = \frac{RT}{(V-b)} - \frac{a}{V^2} \tag{1.1}$$

This was a modification to the Ideal Gas relationship proposed by Clapeyron in 1834.[2]

$$Z = \frac{PV}{RT} = 1 \tag{1.2}$$

Here, P is pressure, V is specific volume, T is absolute temperature, and R is the ideal gas constant. The term, Z, is called the compressibility and is unitless. For convenience and to eliminate units, we introduce three dimensionless ratios.

$$\begin{aligned} P_R &= \frac{P}{P_C} \\ V_R &= \frac{V}{V_C} \\ T_R &= \frac{T}{T_C} \end{aligned} \tag{1.3}$$

The subscript C indicates the critical point and R indicates reduced (or non-dimensionalized). The critical point is where (temperature, pressure, and density or specific volume) of the saturated liquid and saturated vapor are indistinguishable. All stable chemical substances exhibit a critical point. Some molecules break apart or otherwise restructure at elevated pressures or temperatures so that they never reach what might be their critical point. We will only be concerned with fluids that are chemically stable and exhibit a critical point. Other than water (steam) we will not discuss particular fluids. We can non-dimensionalize the van der Waals EOS:

$$Z_C P_R = \frac{T_R}{(V_R - B)} - \frac{A}{V_R^2} \tag{1.4a}$$

where A and B are non-dimensionalized versions of a and b from Equation 1.1.

$$A = \frac{a}{RT_C V_C} \tag{1.4b}$$

$$B = \frac{b}{V_C} \tag{1.4c}$$

The van der Waals EOS is simple and so is often used for the purposes of illustration. In this first figure we see the reduced pressure, P_R, plotted against the reduced specific volume, V_R.

---

[1] Johannes Diderik van der Waals (1837–1923) Dutch theoretical physicist and thermodynamicist.
[2] Benoît Paul Émile Clapeyron (1799–1864) French engineer and physicist.

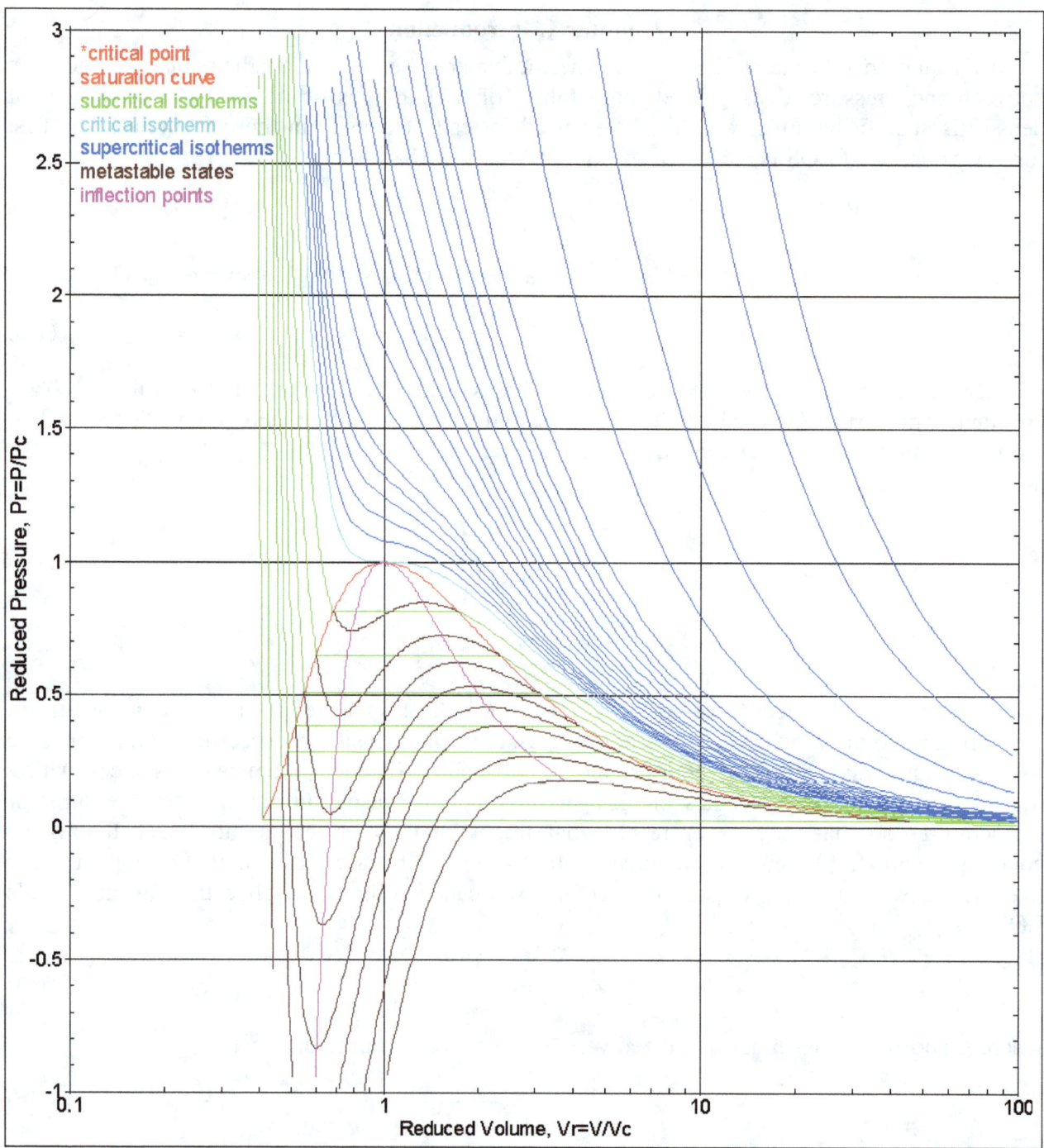

Figure 3. van der Waals Equation of State

The critical point is at the top of the red curve where the magenta and cyan curves also meet. The red curve runs through all of the saturated liquid (on the left) and saturated vapor (on the right) states. The green, cyan, and blue curves are isotherms (lines of constant temperature). The cyan curve is the critical temperature. The brown curves show the metastable states. These are not in equilibrium and will not persist; however, they do exist (briefly) and are an important feature of fluid behavior, impacting boiling and condensation.

Note that there are three locations along the horizontal (specific volume) axis where the green/brown curves intersect any given value on the vertical (pressure) axis. These correspond to the saturated liquid, the point with no formal name, and the saturated vapor. The first (left/liquid) and third (right/vapor) of these intersections coincide with the red saturation curve. The magenta curve passes through the lowest and highest points on the brown metastable curves. These inflection points are called *spinodals*. The area under the red curve is called the *vapor dome*. It is often rightly argued that the first and second derivative of the pressure curves at the critical point (top of the red curve) must be zero in order to assure this shape (maximum).

$$\frac{\partial P}{\partial V} = \frac{\partial^2 P}{\partial V^2} = 0 \tag{1.5}$$

An interesting observation is that the same is implied by three coincident (equal) roots of the pressure equation at this point; that is, all three roots are equal to V_C; thus, the interest in the three roots. As has been illustrated elsewhere[3], there should be exactly three (three and only three) roots of the pressure equation along any isotherm. Furthermore, the area on the left side of the unnamed central root must equal that on the right side. This is called Maxwell's Criterion.[4]

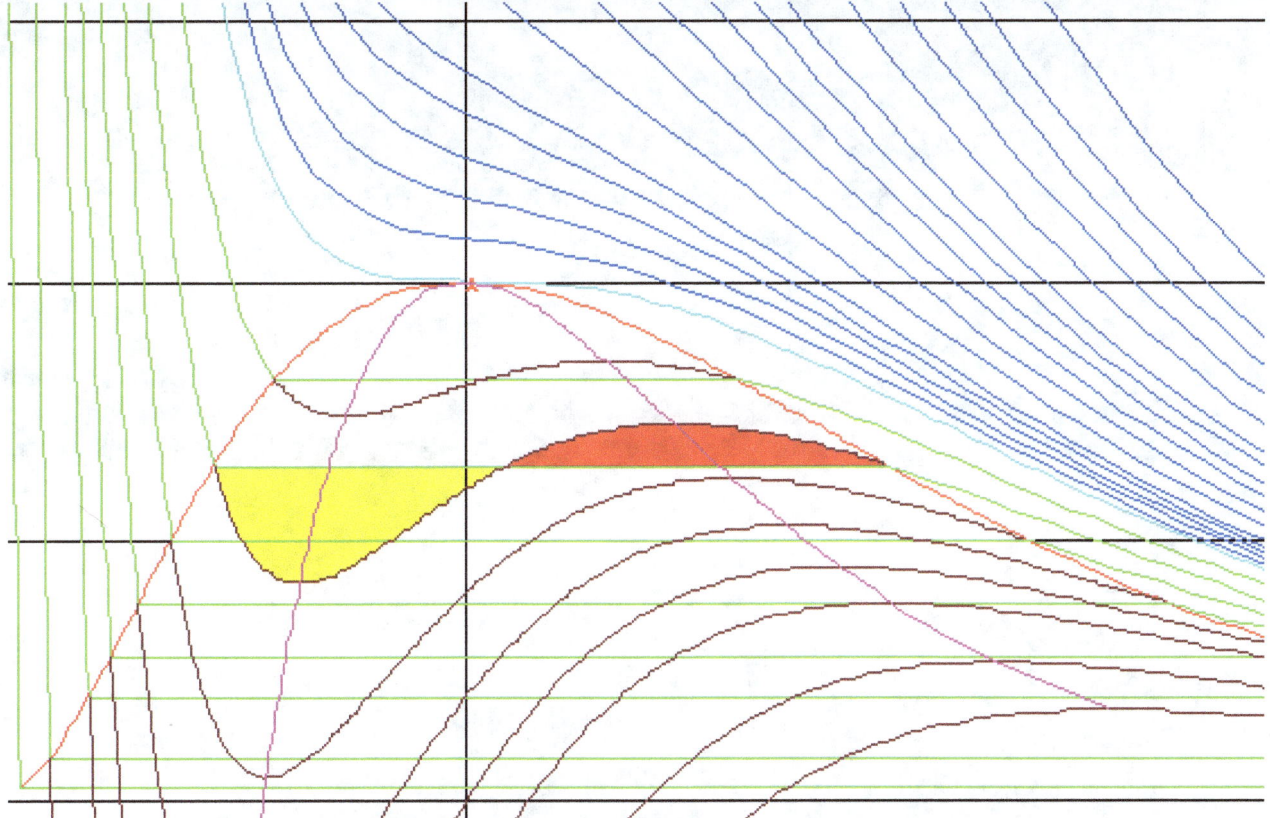

Figure 4. Maxwell's Criterion of Equal Areas

This follows directly from Maxwell's Relationships and the following integral:

$$\int_{v_f}^{v_g} p \, dv = p_{SAT}(v_g - v_f) \tag{1.6}$$

Having the right shape and satisfying Maxwell's Criterion are just two of the many things to consider when evaluating an EOS. That is the reason for this text and why so many figures have been collected so that the reader can make such an evaluation. We will now consider the various figures and a list of common equations.

---

[3] Benton, D. J., *Thermodynamic and Transport Properties of Fluids*, Amazon, 2019.
[4] James Clerk Maxwell (1831–1879) Scottish scientist, physicist, and mathematician.

Chapter 2. Pressure vs. Volume

The fundamental equation for most equations of state is pressure vs. volume. The van der Waals EOS is expressed by Equation 1.1. In order to evaluate the accuracy of this and other equations of state, we consider the empirical data of Nelson and Obert.[5] This is not a formulation per se; however, I have gone to considerable trouble to digitizing the original curves and implementing these computationally for the purposes of illustration. The implementation can be found on the website listed beneath the Preface.

The second EOS we will consider is that of Boltzmann:[6]

$$Z_C P_R = \frac{T_R}{V_R}\left(1 + \frac{B}{V_R} + \frac{5B^2}{V_R^2}\right) - \frac{A}{V_R^2} \tag{2.1}$$

The third EOS we will consider is that of Berthelot:[7,8]

$$Z_C P_R = \frac{T_R}{(V_R - B)} - \frac{A}{T_R V_R^2} \tag{2.2}$$

The fourth EOS we will consider is that of Dieterici:[9]

$$Z_C P_R = \frac{T_R}{\left(V_R - \frac{1}{2}\right) e^{\frac{2}{T_R V_R}}} \tag{2.3}$$

The fifth EOS we will consider is that of Clausius:[10]

$$Z_C P_R = \frac{T_R}{(V_R - B)} - \frac{A}{T_R (V_R + C)^2}$$
$$A = \frac{(1+C)^3}{(1-B)^2}$$
$$C = \frac{3(1-B)}{2} - 1 \tag{2.4}$$
$$Z_C = \frac{1}{(1-B)} - \frac{A}{(1+C)^2}$$

---

[5] Nelson, L. C. and Obert, E. F., "Generalized PVT Properties of Gases," *Transactions of the ASME*, Vol. 76, pp. 1057–1066, 1954.
[6] Ludwig Eduard Boltzmann (1844–1906) Austrian physicist and philosopher, father of statistical mechanics.
[7] Daniel Berthelot (1865-1927) French biologist and physicist, professor and researcher at the Academy of Sciences and the Academy of Medicine.
[8] D. Berthelot, "Travaux et Mémoires du Bureau International des Poids et Mesures," Vol. 13, 1907.
[9] C. Dieterici, (German) *Annals of Physical Chemistry* (Wiedemann's Annalen der Physik und Chemie), Vol. 69, p. 685, 1899.
[10] Rudolf Julius Emanuel Clausius (1822–1888) German physicist and mathematician.

The sixth EOS we will consider is Abbott's modification to the Clausius EOS:[11]

$$A = \frac{27}{64 Z_C} \left(\frac{1 - \frac{49}{60} w}{T_R^{0.6}} + \frac{49}{60} \frac{w}{T_R^{2.6}} \right) \quad (2.5)$$

where w is the Pitzer acentric factor:[12]

$$w = -\log_{10}\left(P_R^{sat}\right) - 1 \quad (2.6)$$
$$at\, T_R = 0.7$$

The seventh EOS we will consider was developed by Redlich & Kwong:[13]

$$Z_C P_R = \frac{T_R}{(V_R - B)} - \frac{A}{V_R(V_R + B)\sqrt{T_R}} \quad (2.7)$$

The eight EOS we will consider is Soave's modification to the R-K EOS:[14]

$$A = 0.42748 \left[1 + \left(0.480 + 1.574 w - 0.176 w^2\right)\left(1 - \sqrt{T_R}\right) \right]^2 \quad (2.8)$$

The ninth EOS we will consider is Fuller's modification to the R-K EOS:[15]

$$\frac{A}{A_C} = \frac{1 - \frac{49}{60} w}{T_R^{0.6}} + \frac{49}{60} \frac{w}{T_R^{2.6}} \quad (2.9)$$

The tenth EOS we will consider is the Peng-Robinson:[16]

$$Z_C P_R = \frac{T_R}{(V_R - B)} - \frac{A \alpha}{\left(V_R^2 + 2 B V_R - B^2\right)} \quad (2.10a)$$

$$A = \frac{0.45724}{Z_C} \quad (2.10b)$$

$$B = \frac{0.0778}{Z_C} \quad (2.10c)$$

$$\alpha = \left(1 + \kappa\left(1 - Tr^{\frac{1}{2}}\right)\right)^2 \quad (2.10d)$$

$$\kappa = 0.37464 + 1.54226 \omega - 0.26992 \omega_2 \quad (2.10d)$$

---

[11] Abbott, M. M., "Cubic Equations of State," *AIChE Journal*, Vol. 19, p. 596, 1973.

[12] Pitzer, K. S., et al., "Volumetric and Thermodynamic Properties of Fluids II: Compressibility Factor, Vapor Pressure, and Entropy of Vaporization," *Journal of the American Chemical Society*, Vol. 77, pp. 3433, 1955.

[13] Redlich, O. and Kwong, J. N. S., "On The Thermodynamics of Solutions," *Chemical Review*, Vol. 44, No. 1, pp. 233–244, 1949.

[14] Soave, G., "Equilibrium Constants from a Modified Redlich-Kwong Equation of State," *Chemical Engineering Science*, Vol. 27, No. 6, pp. 1197–1203, 1972.

[15] Fuller, G. G., "A Modified Redlich-Kwong-Soave Equation of State Capable of Representing the Liquid State," *Industrial & Engineering Chemistry Fundamentals*, Vol. 15, No. 4, pp. 254-257, 1976.

[16] Peng, D. Y. and Robinson, D. B., "A New Two-Constant Equation of State," *Industrial and Engineering Chemistry: Fundamentals*, Vol. 15, pp. 59–64, 1976.

The eleventh formulation we will consider is this Author's modification to the cubic EOS:

$$Z_C P_R = \frac{T_R}{(V_R - B)} - \frac{A}{\left(V_R^2 + CV_R - D^2\right)} \qquad (2.11)$$

where A, B, C, and D are functions of temperature. The remaining formulations are specifically for steam and too complicated to be presented as simple formulas. These are the 1969 version of steam by Keenan, Keyes, Hill, and Moore[17]; the 1984 NBS/NRC version of steam by Haar, Gallagher, and Kell[18]; the 1995 IAPWS version of steam by Wagner and Pruß[19]; and finally the 2022 version of steam by this Author.[20]

In this and the subsequent chapters, we first show the Nelson-Obert generalized relationship, as this is based on actual data and is independent of any particular formulation. This figure is followed by the remaining formulations in the same order. It is important to note in this first set of figures, all of the analytical formulations have meaningful (if not accurate) representation of the metastable states under the vapor dome, while the steam formulations (with the exception of 2020) do not. No formulation for the thermodynamic properties of steam before 2020 ever even tried to properly represent the metastable region. At least Keenan, Keyes, Hill, and Moore discuss this in their Appendix. The others completely ignore it. This region is important for boiling and condensation analyses and has been recognized by Karimi and Leinhard, whose name you will find associated with many publications[21] and more recently by Hurtado, Marro, and Garrido.[22]

---

[17] Keenan, J. H., Keyes, F. G., Hill, P. G., and Moore, J. G., *Steam Tables*, John Wiley & Sons, Inc., 1969.
[18] Haar, L., Gallagher, J. S., and Kell, G. S., *Steam Tables*, NBS/NRC printed by Hemisphere, distributed by McGraw-Hill, 1984.
[19] Wagner, W., and Pruß, A., "The IAPWS Formulation 1995 for the Thermodynamic Properties of Ordinary Water Substance for General and Scientific Use," Journal of Physical Chemistry, Ref. Data 31, pp. 387-535, 2002.
[20] Benton, D. J., *Steam 2020: to 150 GPa and 6000 K*, Amazon, 2020.
[21] Karimi, A., and Lienhard, J. H., "A Fundamental Equation Representing Water in the Stable, Metastable, and Unstable States," EPRI Report No. NP-3328, December, 1983.
[22] Hurtado, P. I., Marro, J., and Garrido, P. L., "Reentrant behavior of the Spinodal Curve in a Nonequilibrium Ferromagnet," Physical Review, Vol. E-70, 2004.

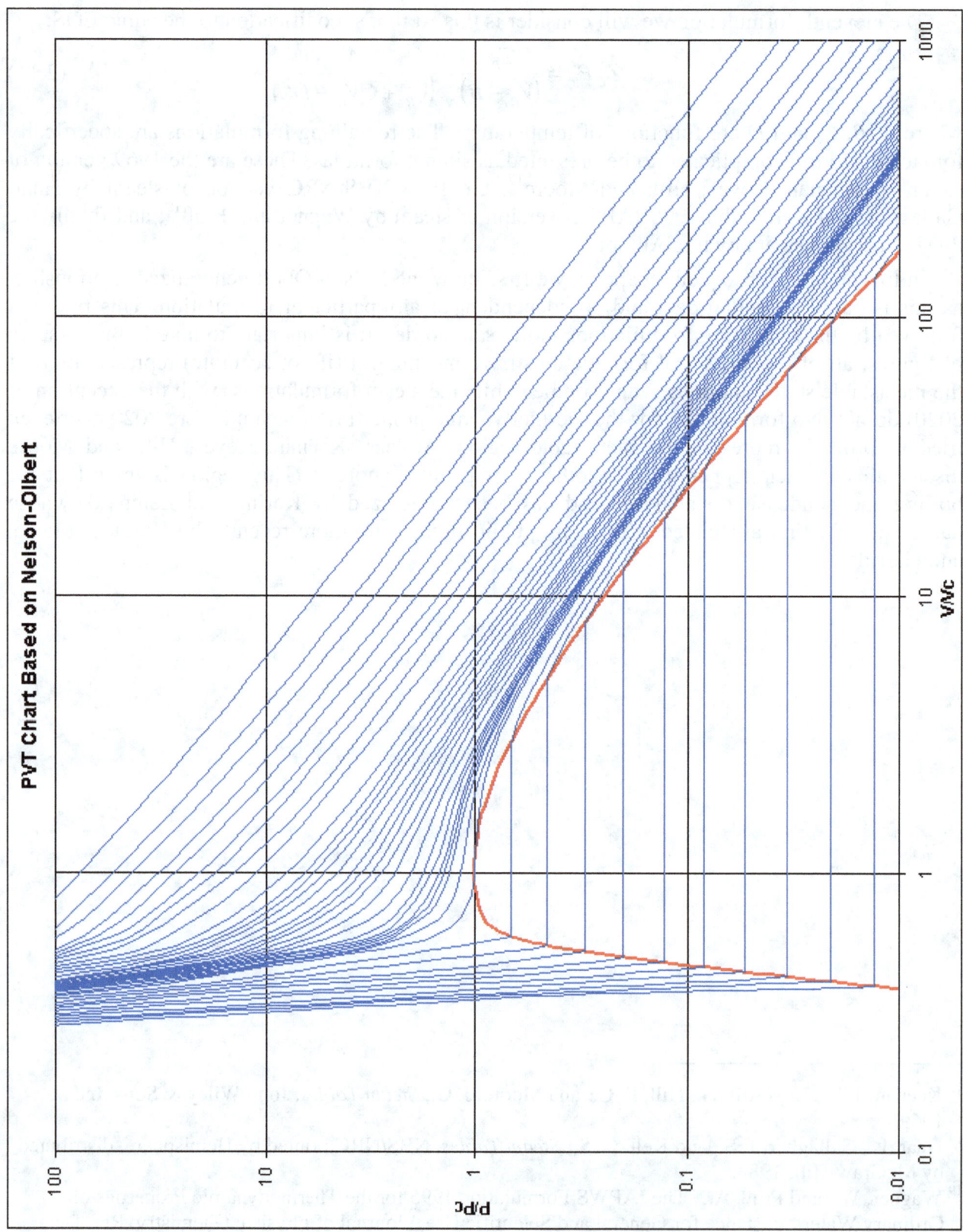

Figure 5. Pr vs. Vr Based on Nelson-Obert

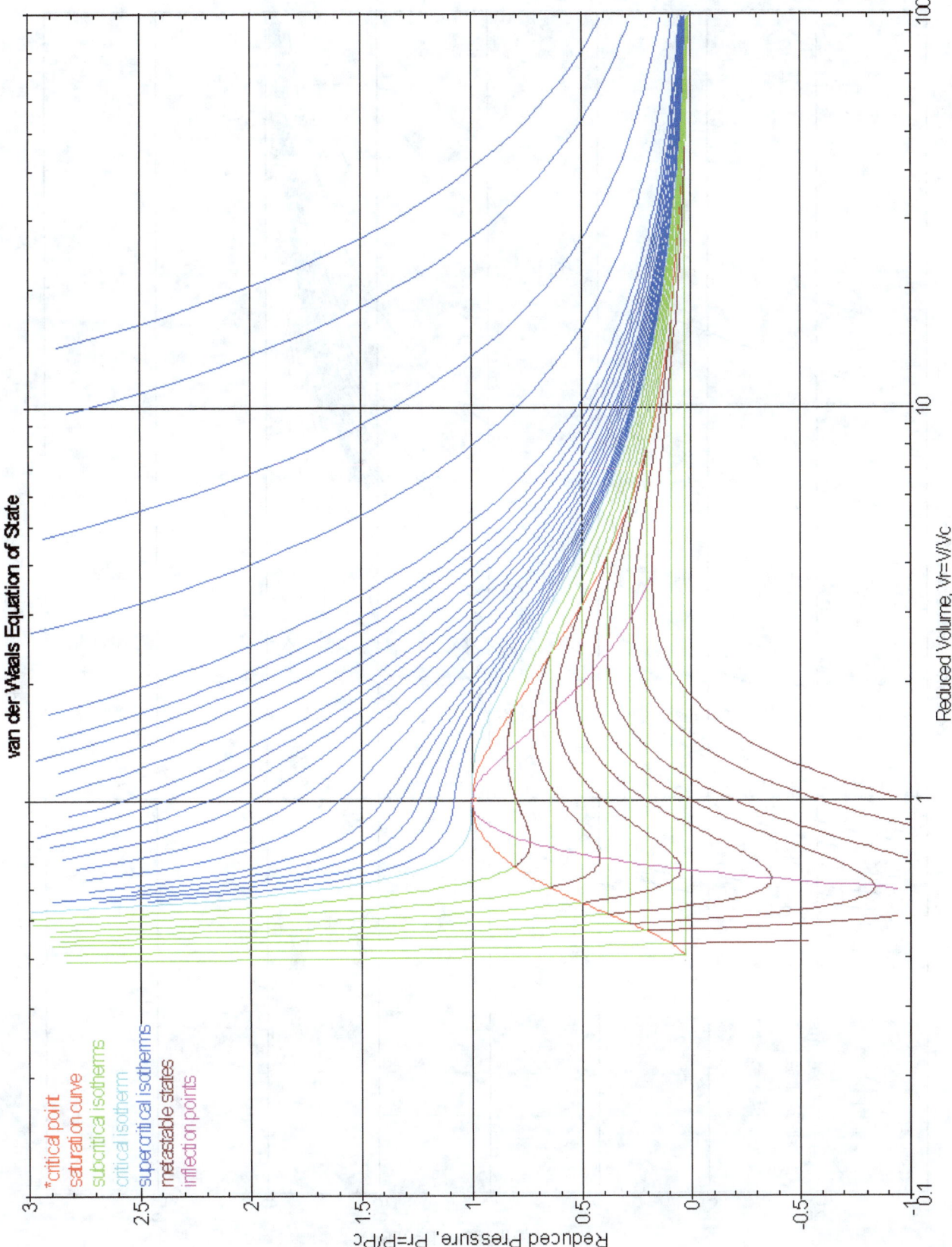

Figure 6. Pr vs. Vr Based on van der Waals

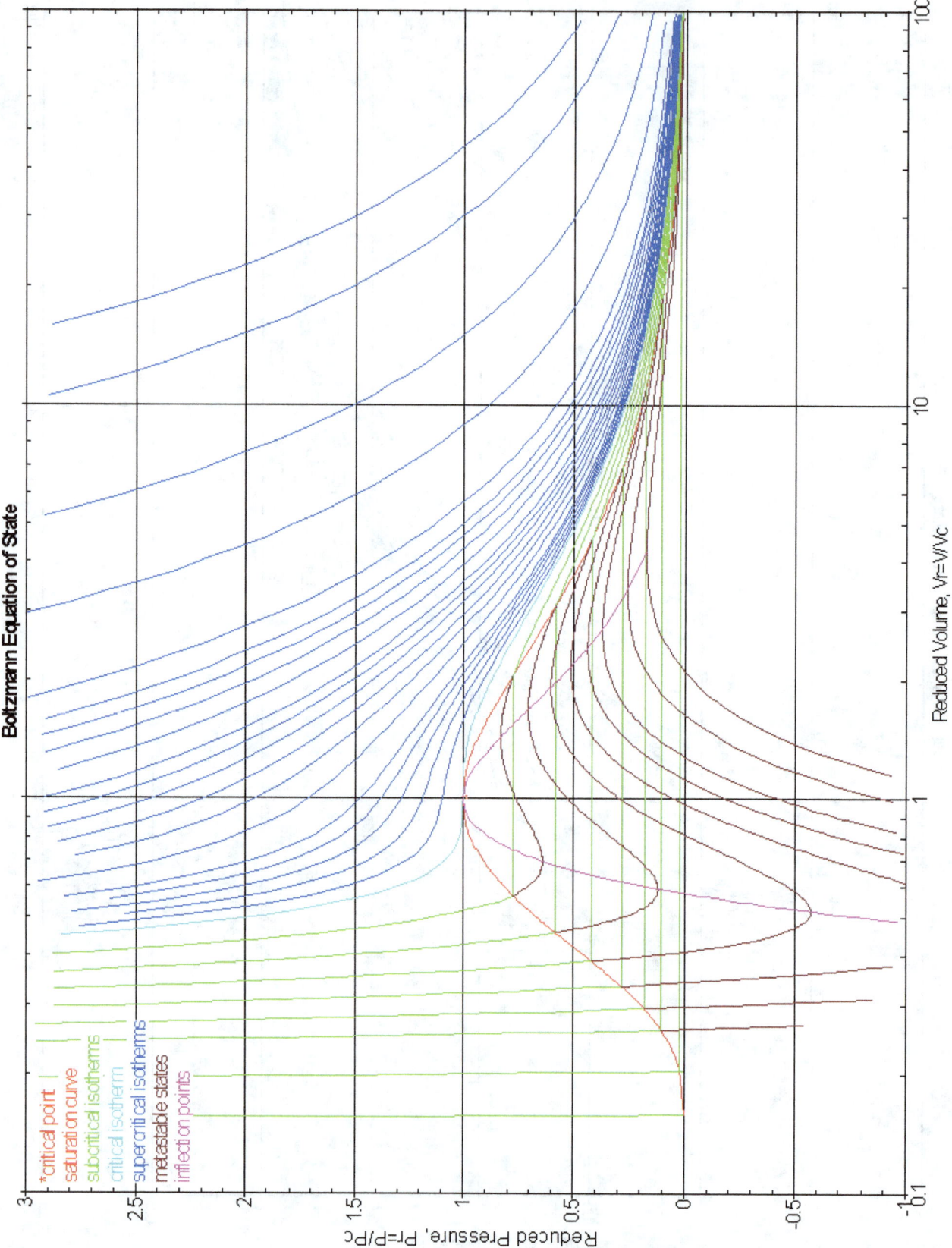

Figure 7. Pr vs. Vr Based on Boltzmann

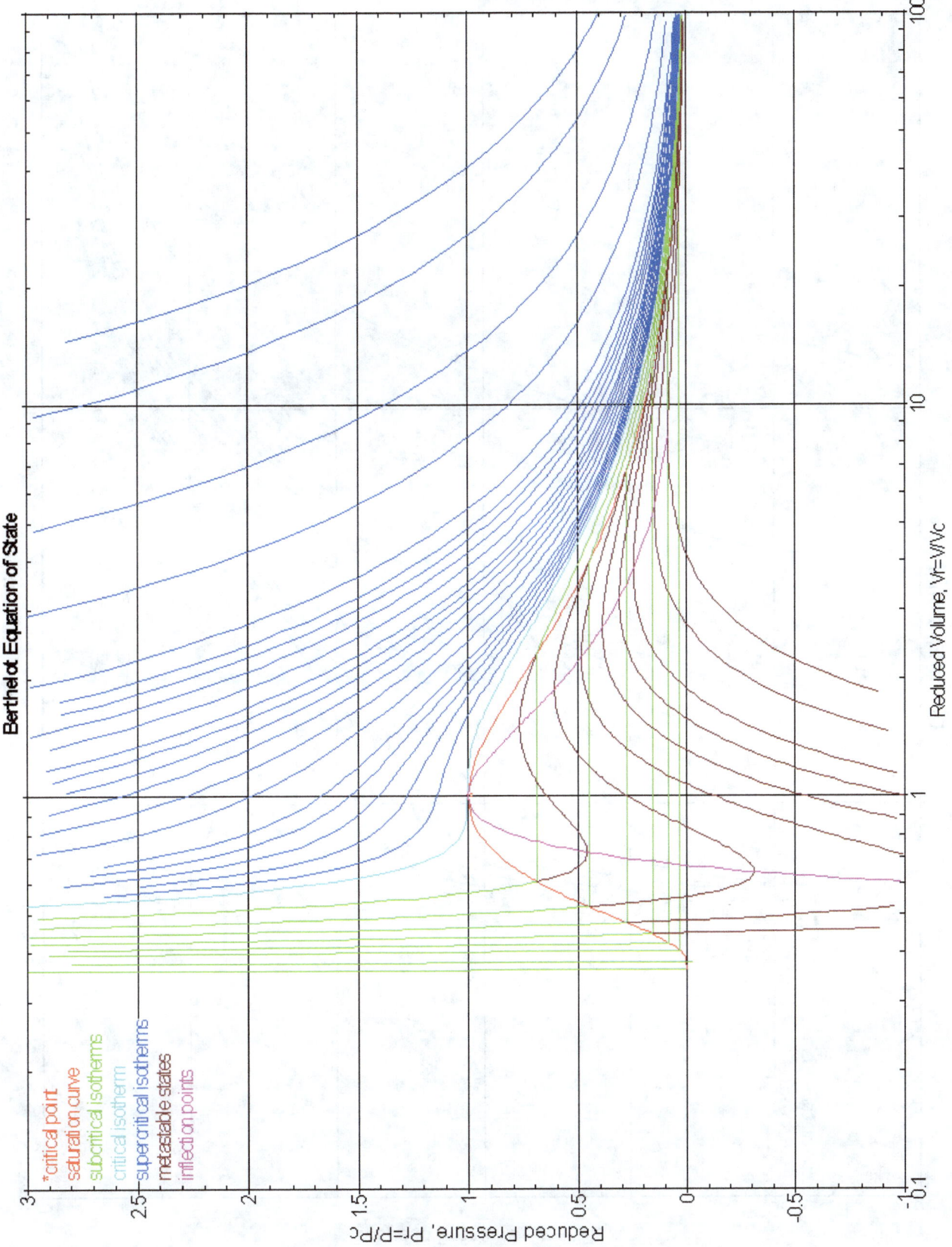

Figure 8. Pr vs. Vr Based on Berthelot

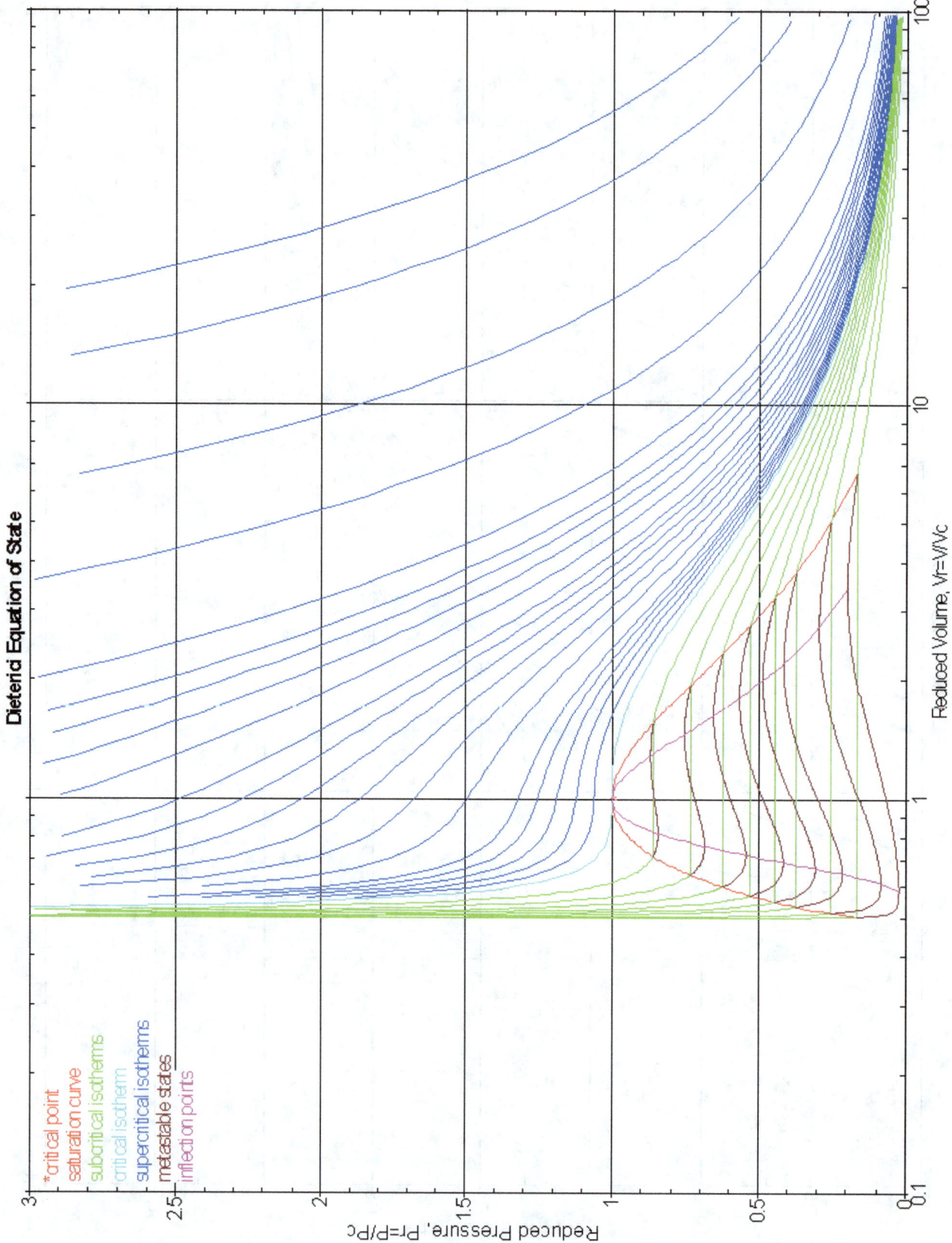

Figure 9. Pr vs. Vr Based on Deterici

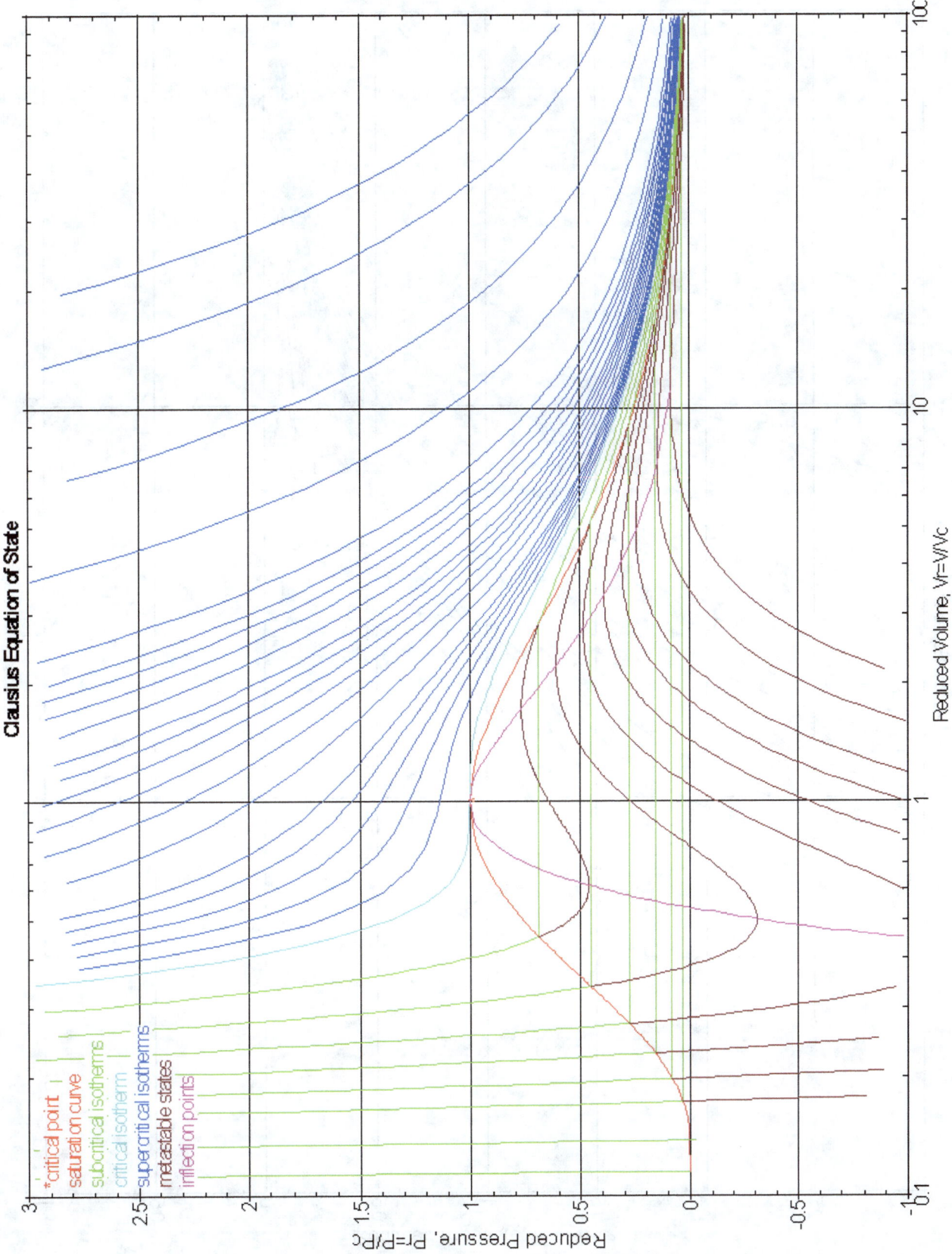

Figure 10. Pr vs. Vr Based on Clausius

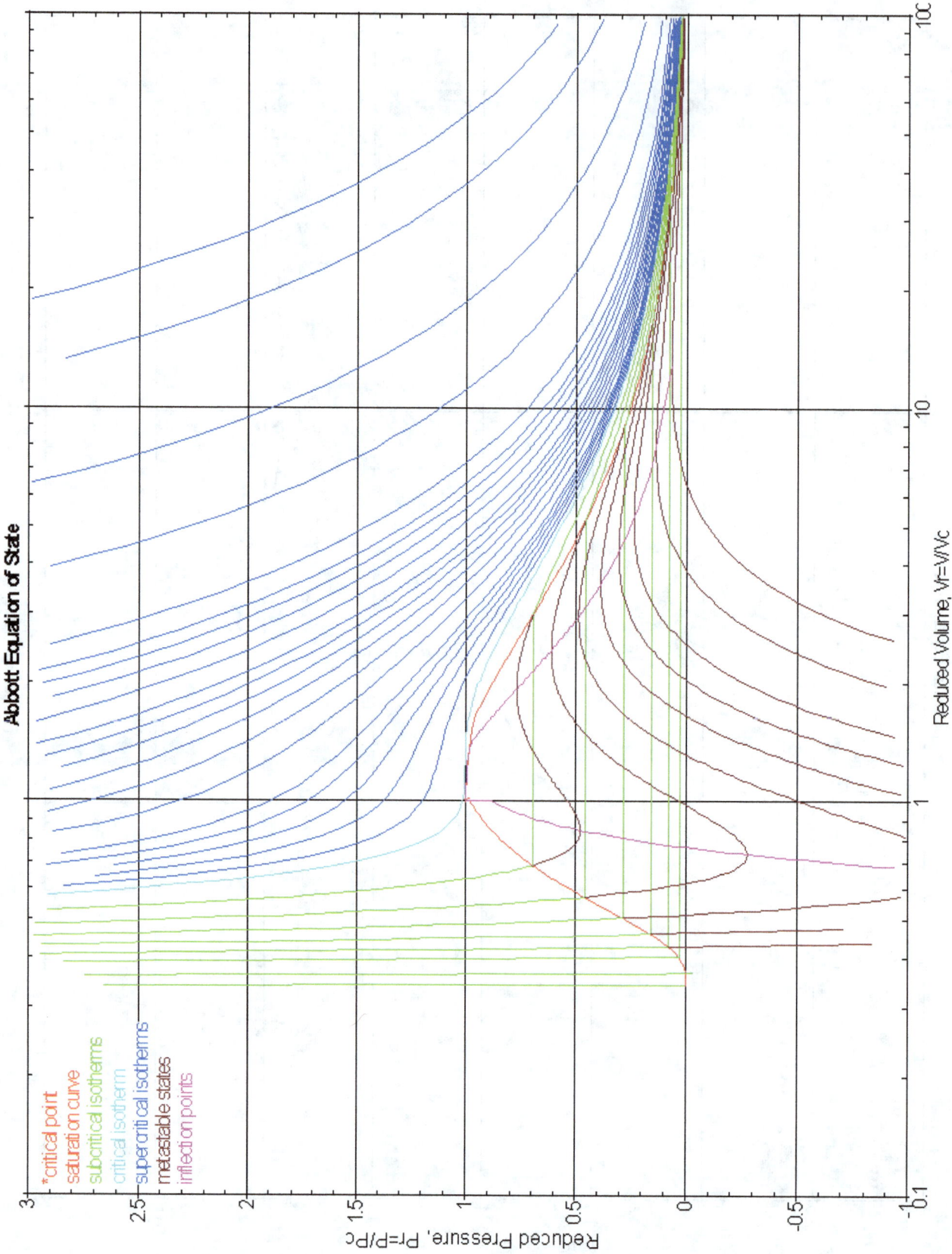

Figure 11. Pr vs. Vr Based on Abbott's Modification

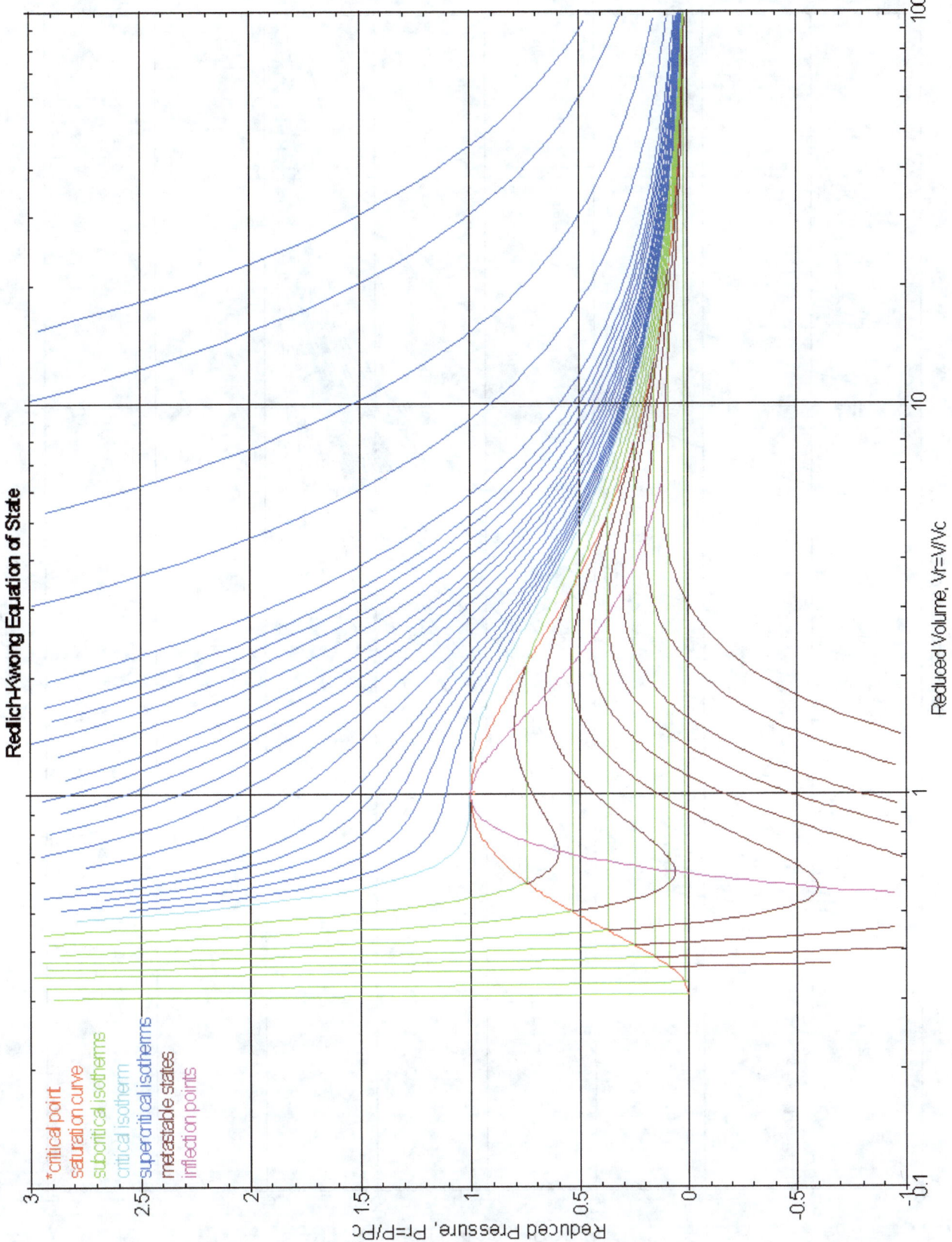

Figure 12. Pr vs. Vr Based on Redlich-Kwong

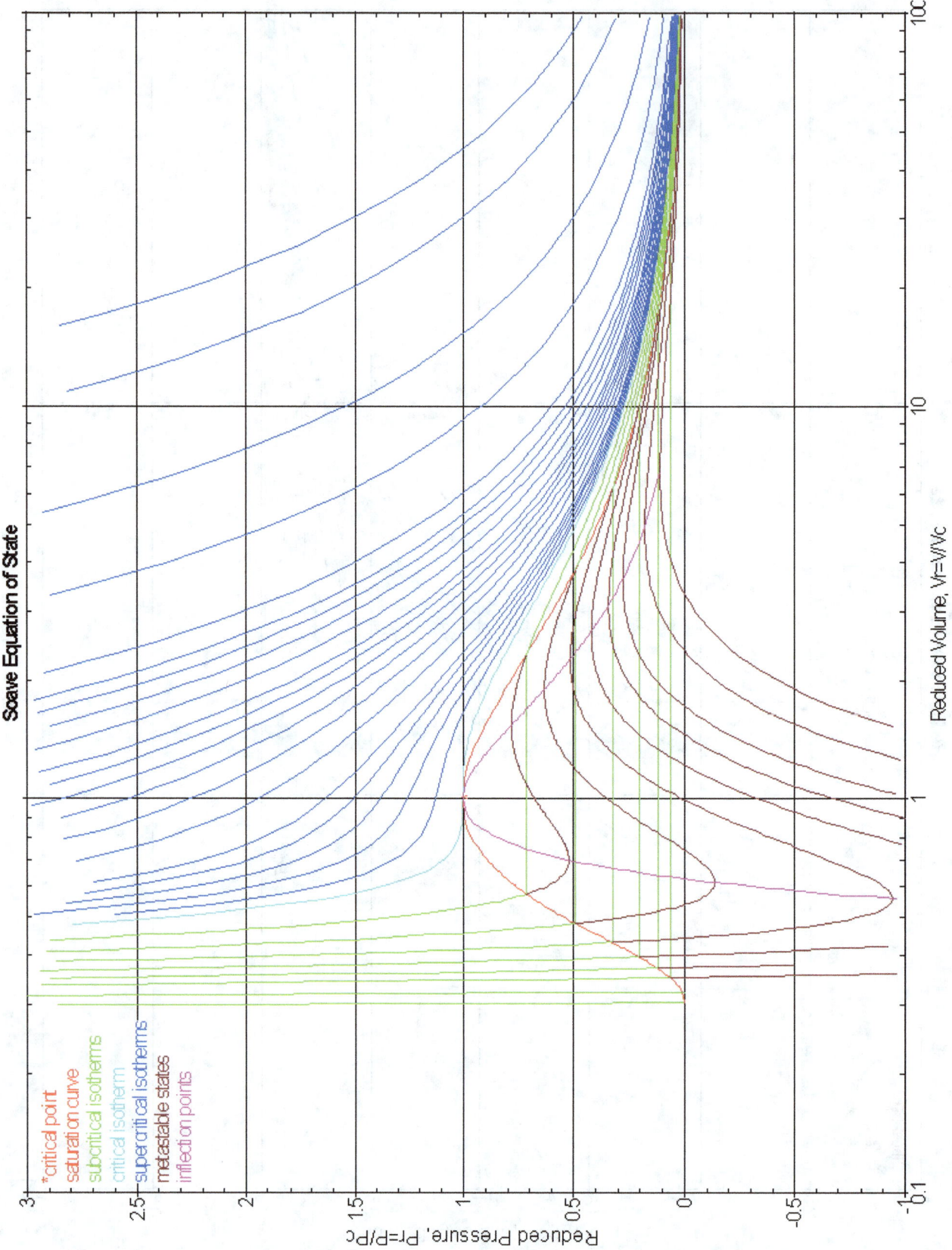

Figure 13. Pr vs. Vr Based on Soave's Modification

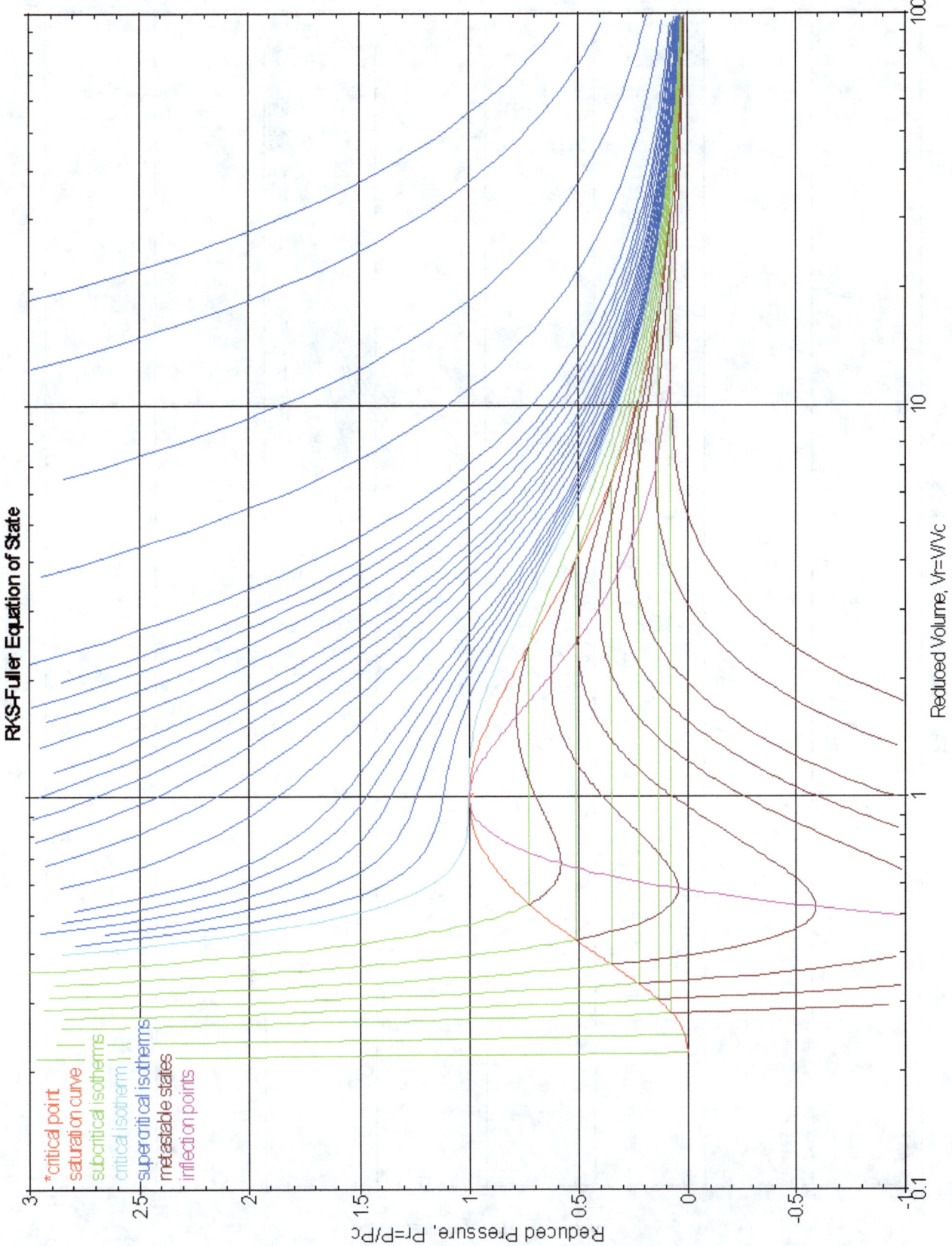

Figure 14. Pr vs. Vr Based on Fuller's Modification

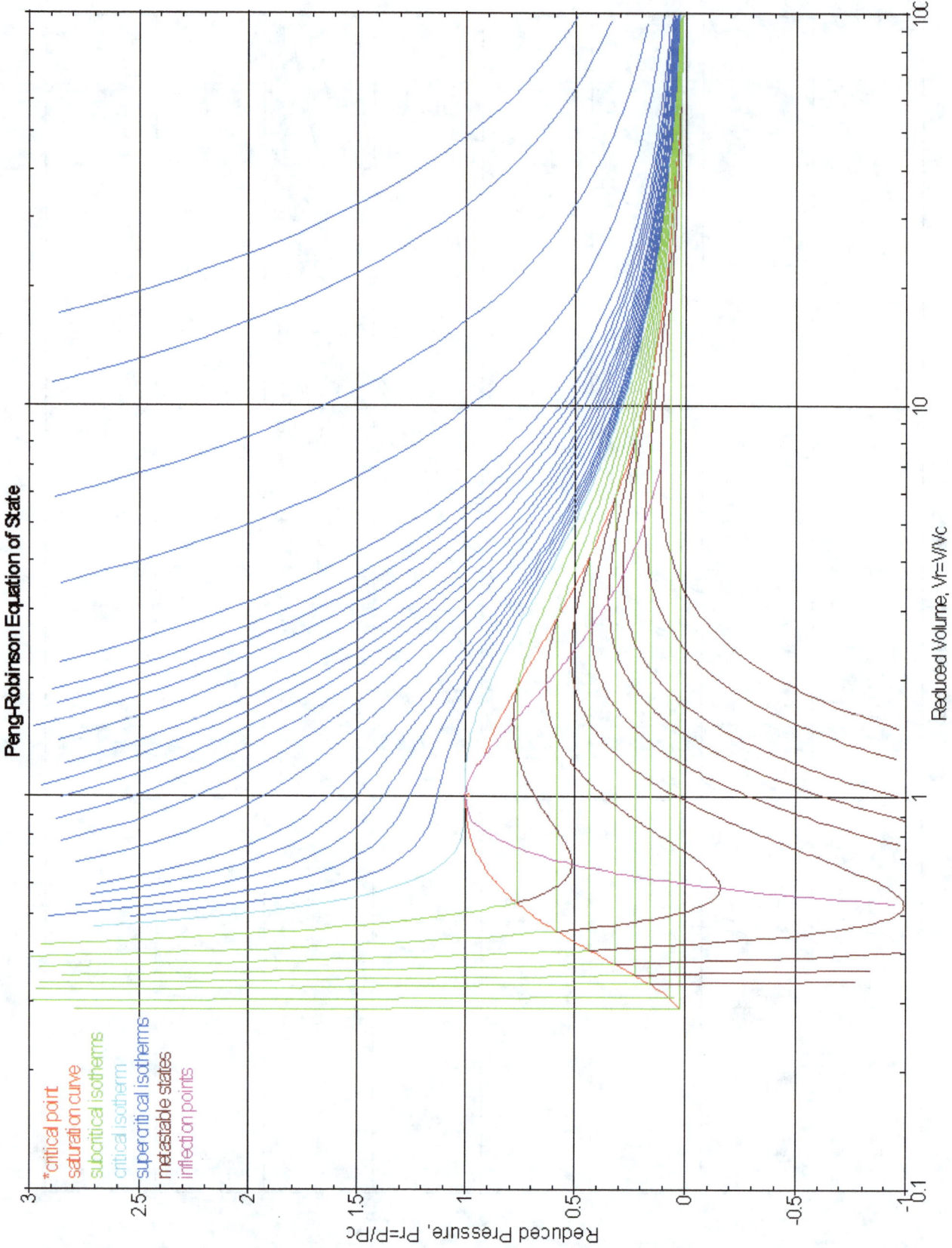

Figure 15. Pr vs. Vr Based on Peng-Robinson

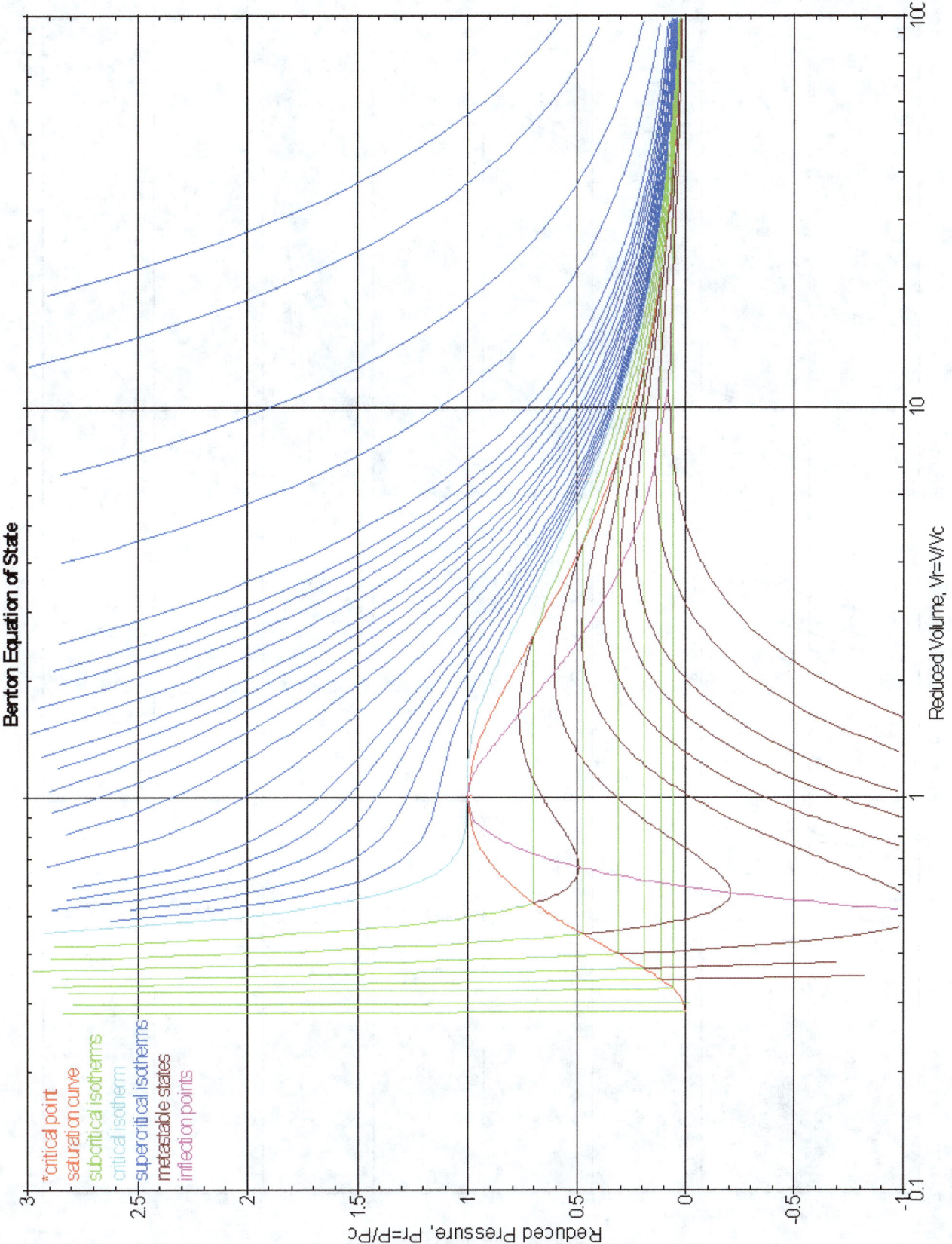

Figure 16. Pr vs. Vr Based on Author's Modification

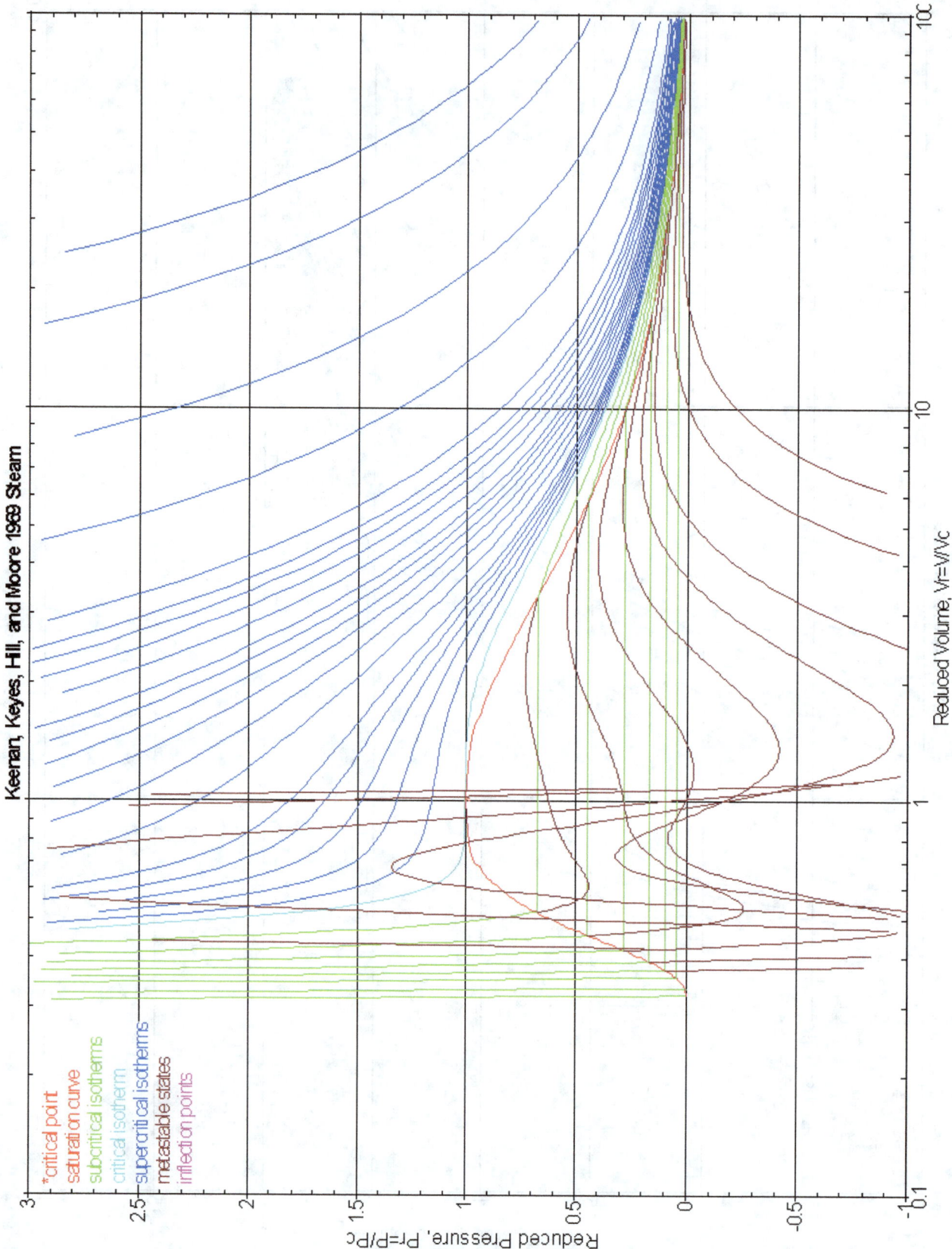

Figure 17. Pr vs. Vr Based on Keenan, Keyes, Hill, and Moore

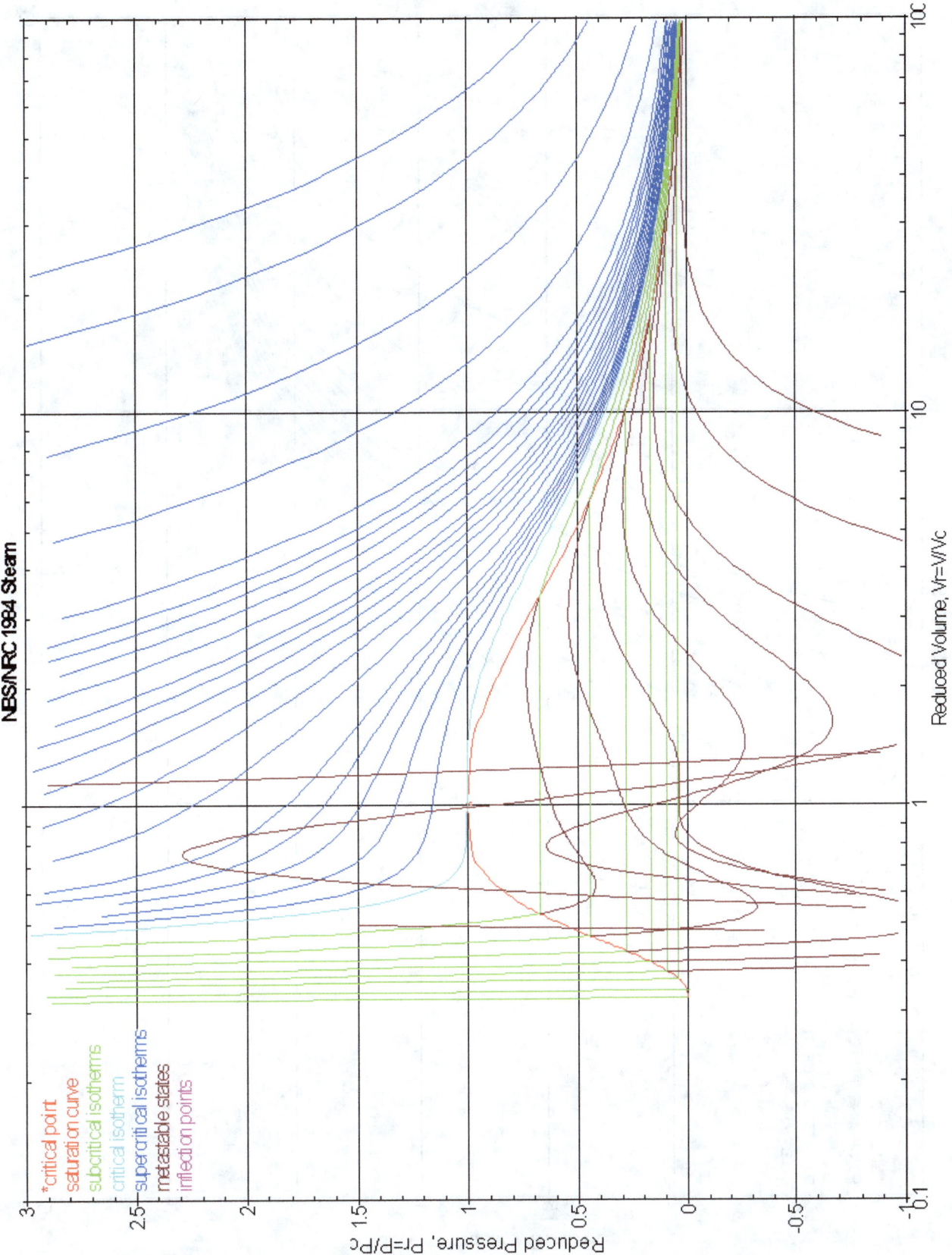

Figure 18. Pr vs. Vr Based on Haar, Gallagher, and Kell

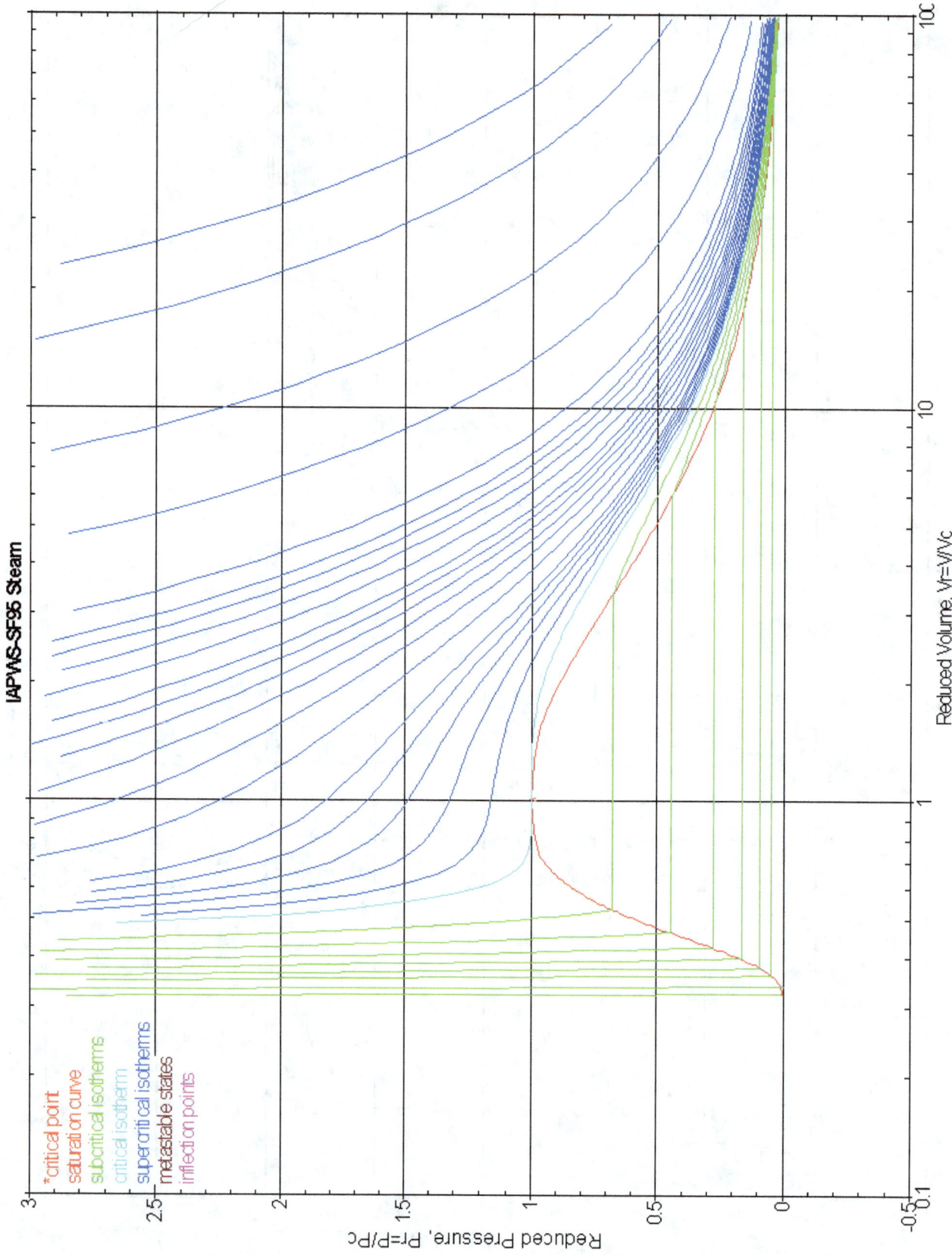

Figure 19. Pr vs. Vr Based on Wagner and Pruß

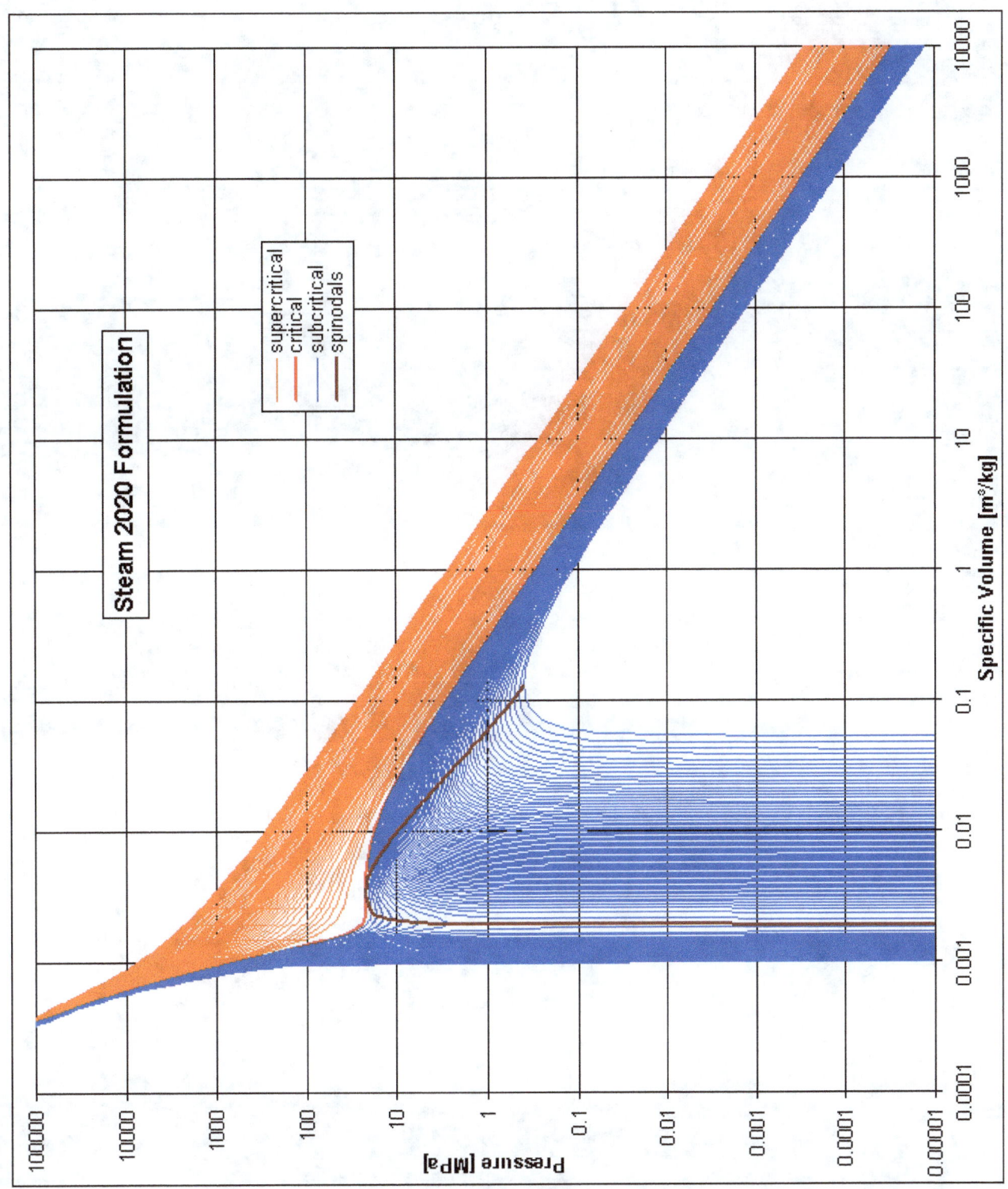

Figure 20. Pr vs. Vr Based on Steam 2020 Formulation

Chapter 3. Temperature vs. Pressure

The same equations of state utilized in the previous chapter are employed here to create isobars (lines of constant pressure) rather than isotherms (lines of constant temperature). This is an implicit process that can be quite involved, as these equations are expressions for pressure as a function of temperature. Again, we begin with the Nelson-Obert data, which has been digitized and approximated so as to facilitate such calculations.

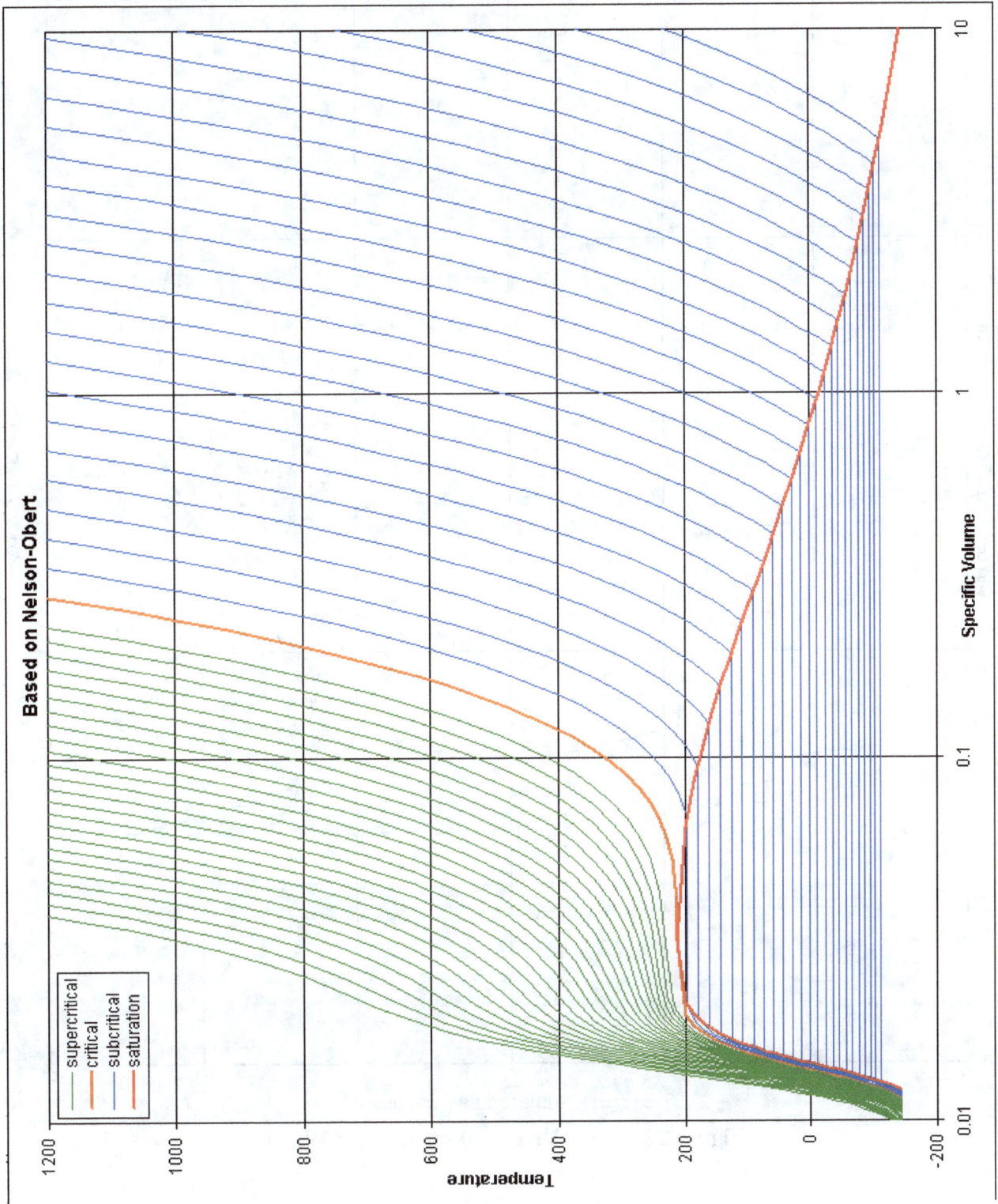

Figure 21. T vs. V Based on Nelson-Obert

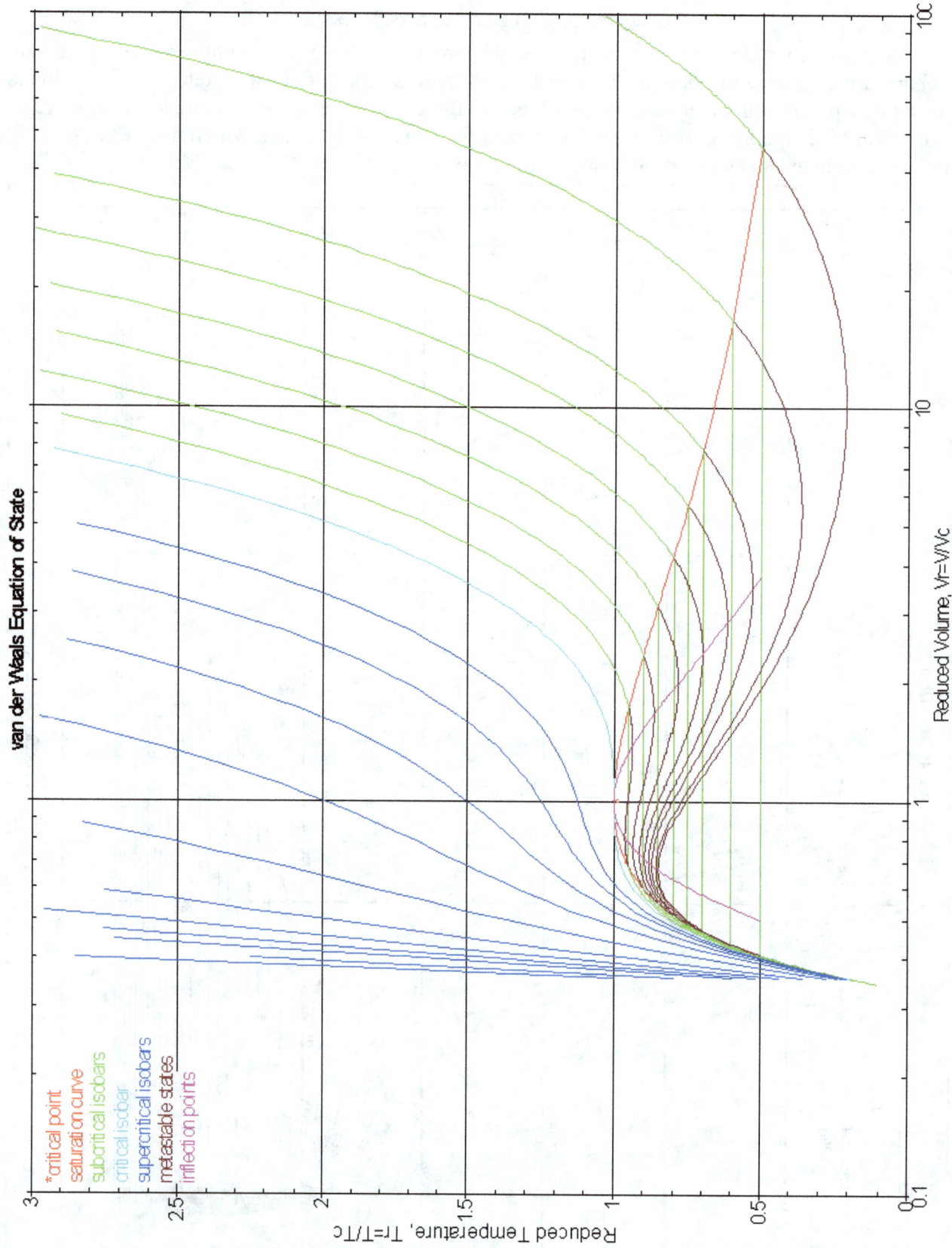

Figure 22. Tr vs. Vr Based on van der Waals

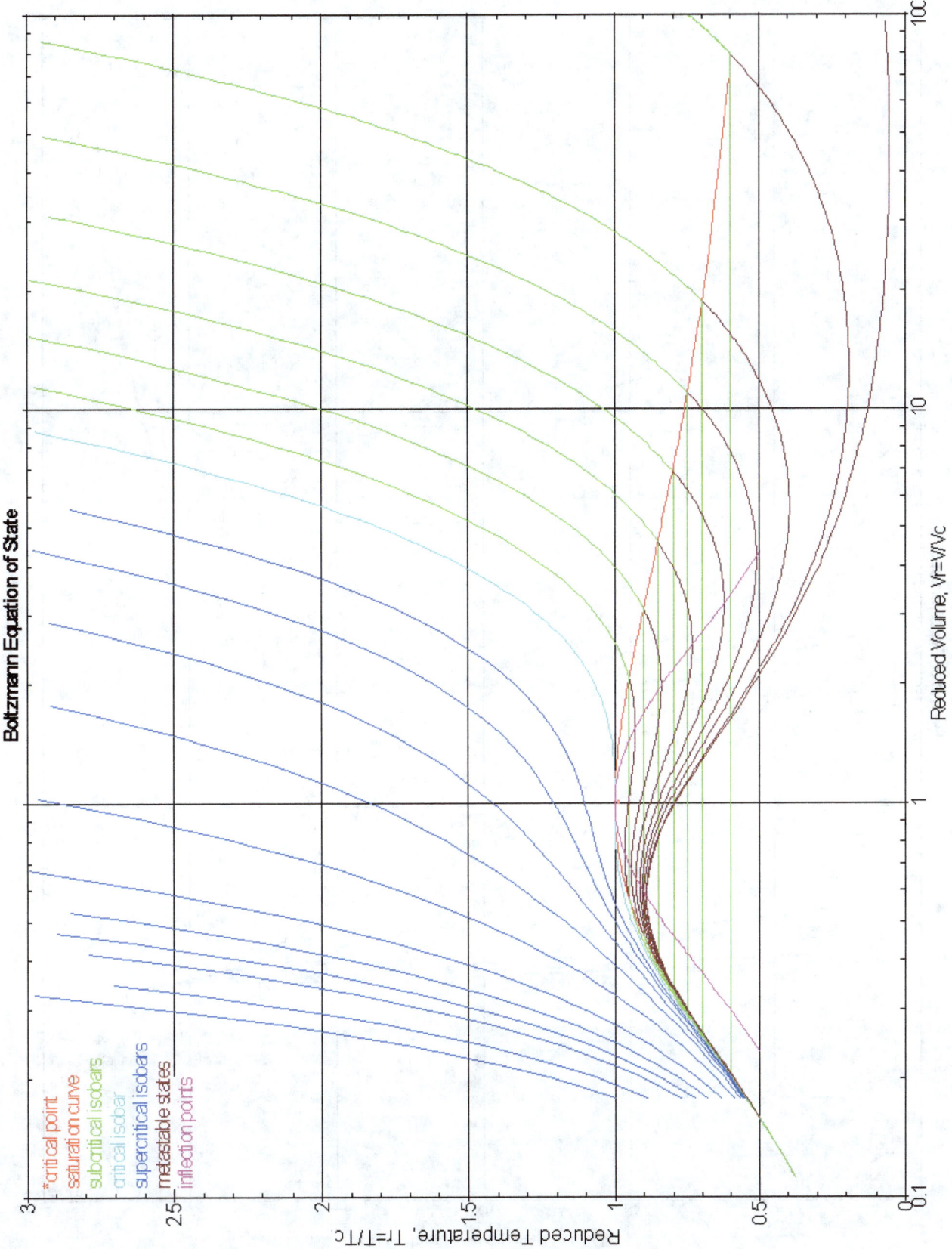

Figure 23. Tr vs. Vr Based on Boltzmann

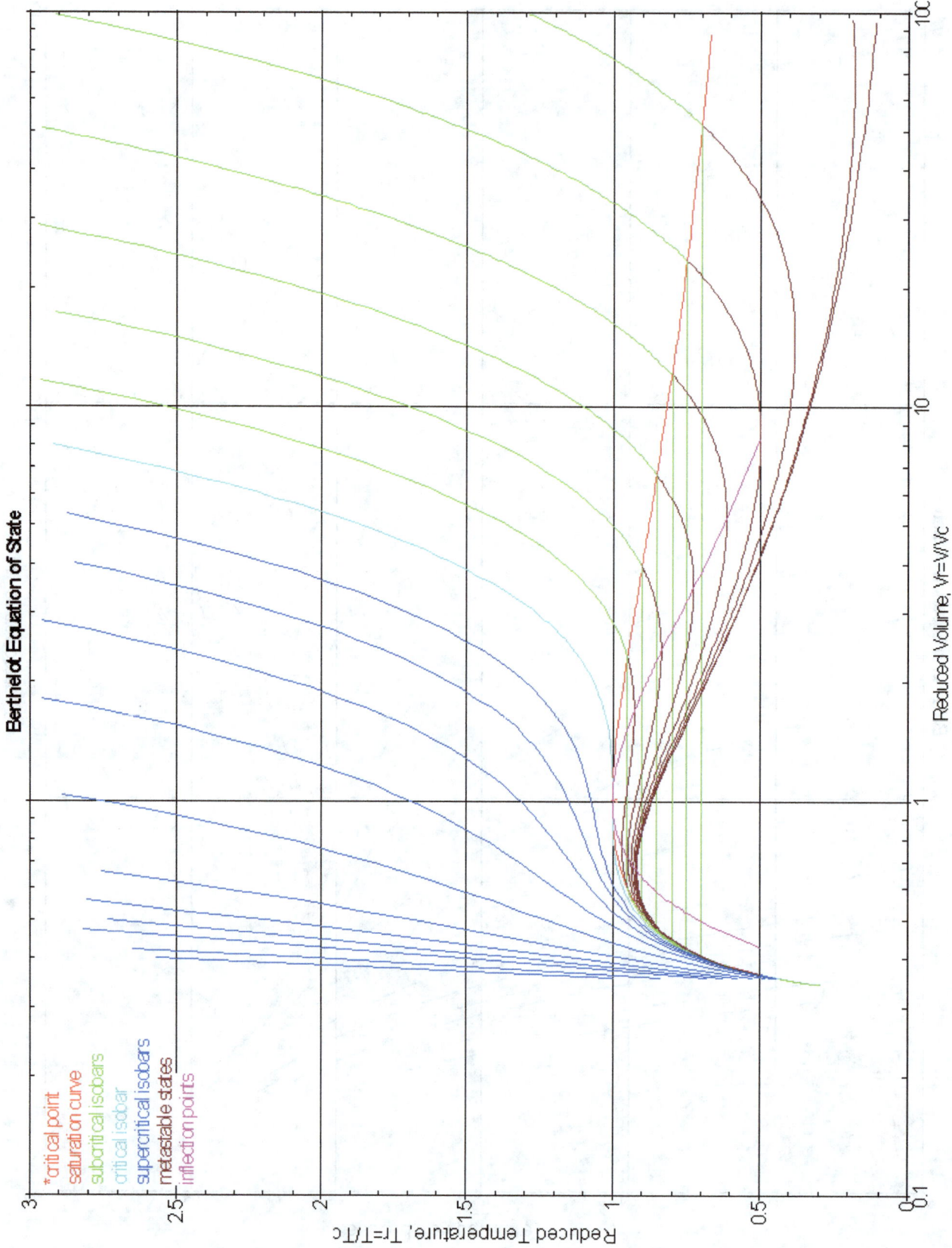

Figure 24. Tr vs. Vr Based on Berthelot

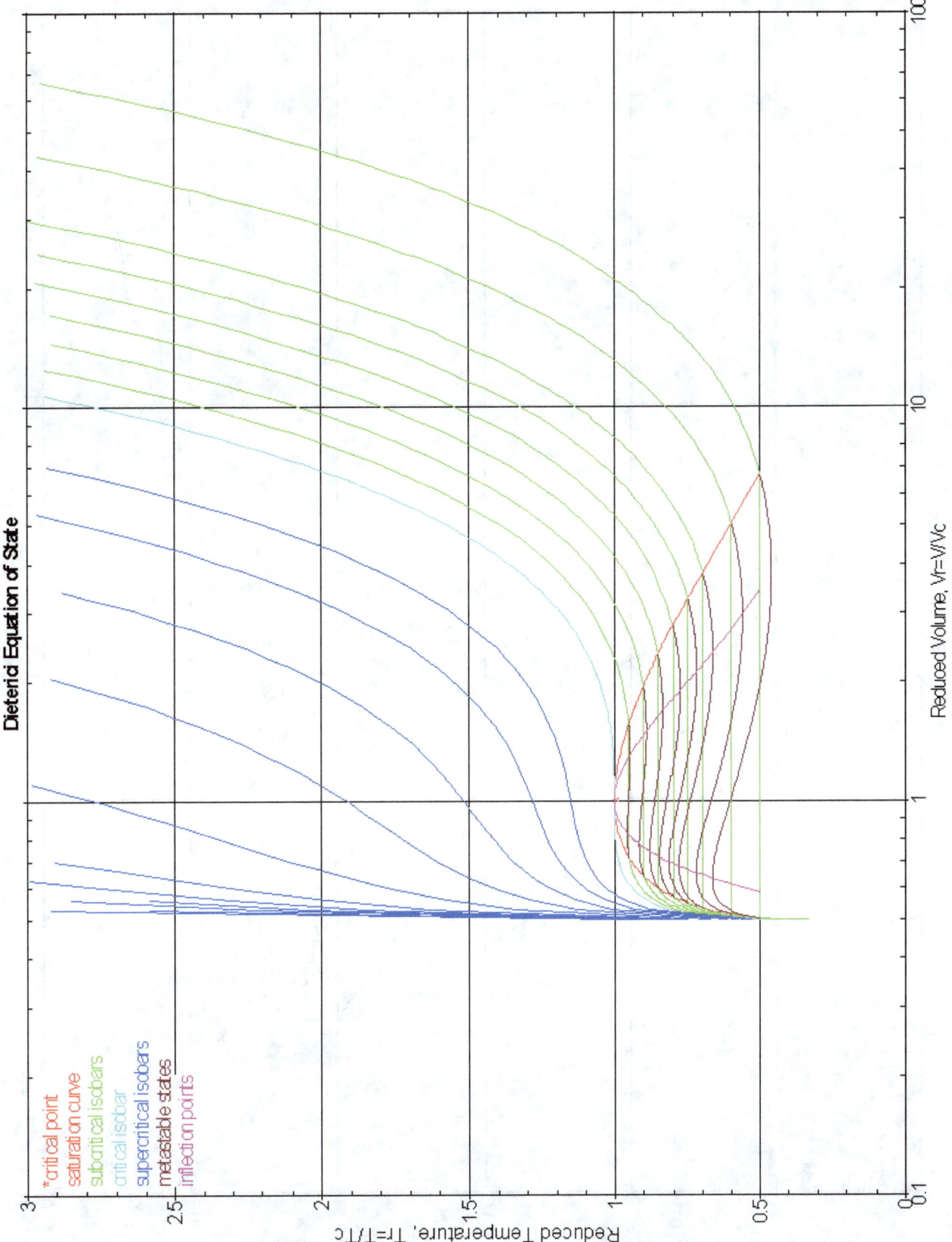

Figure 25. Tr vs. Vr Based on Dieterici

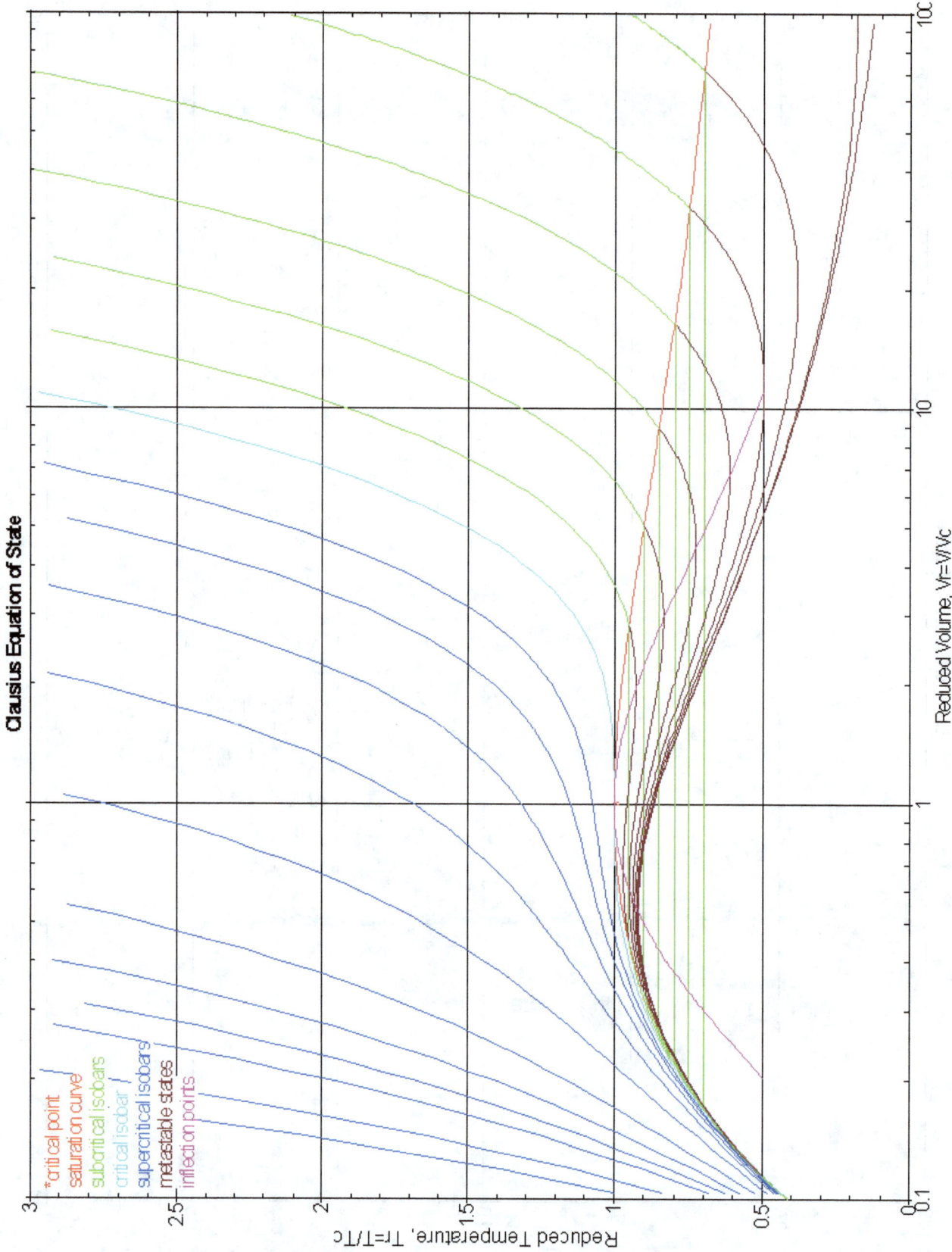

Figure 26. Tr vs. Vr Based on Clausius

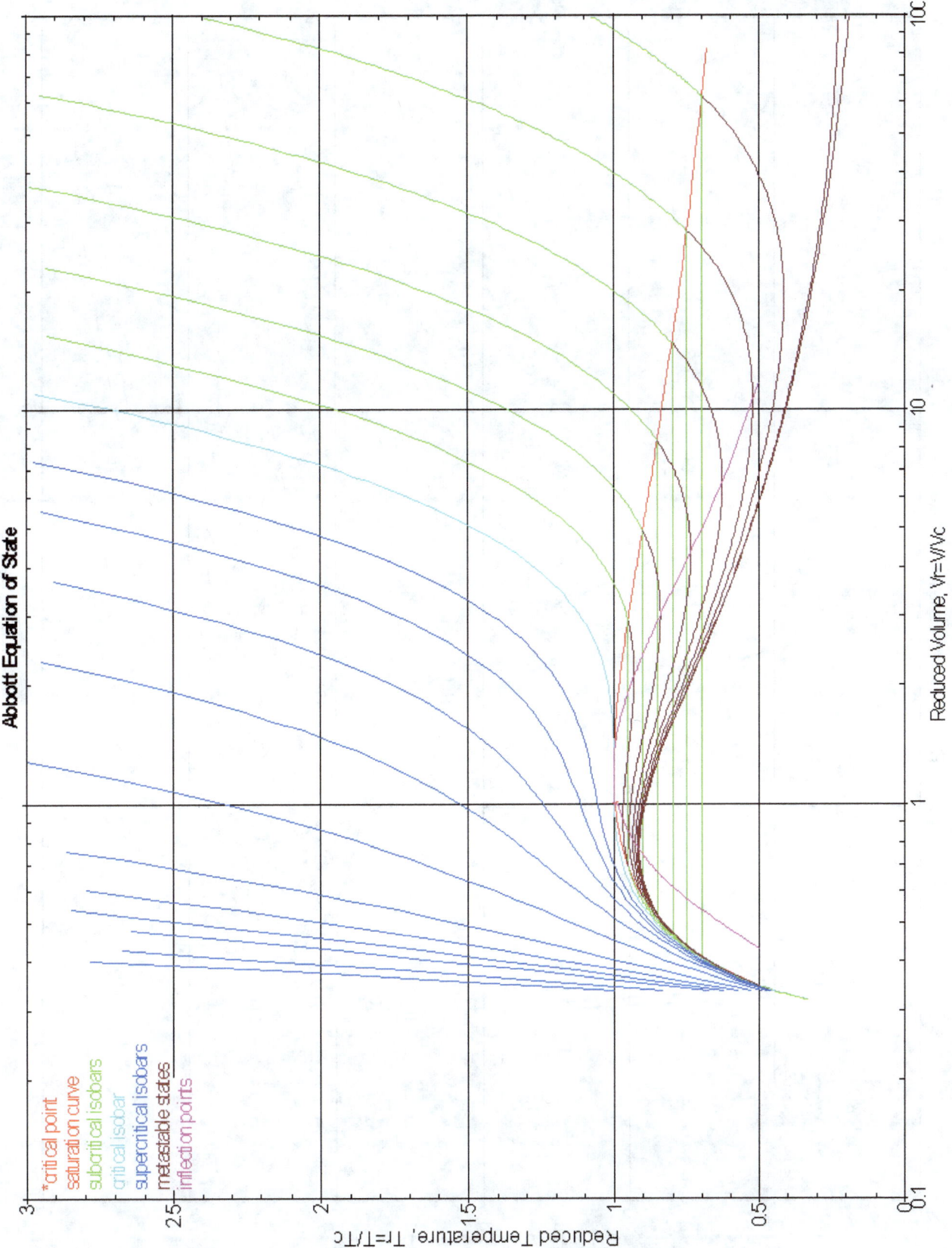

Figure 27. Tr vs. Vr Based on Abbott's Modification

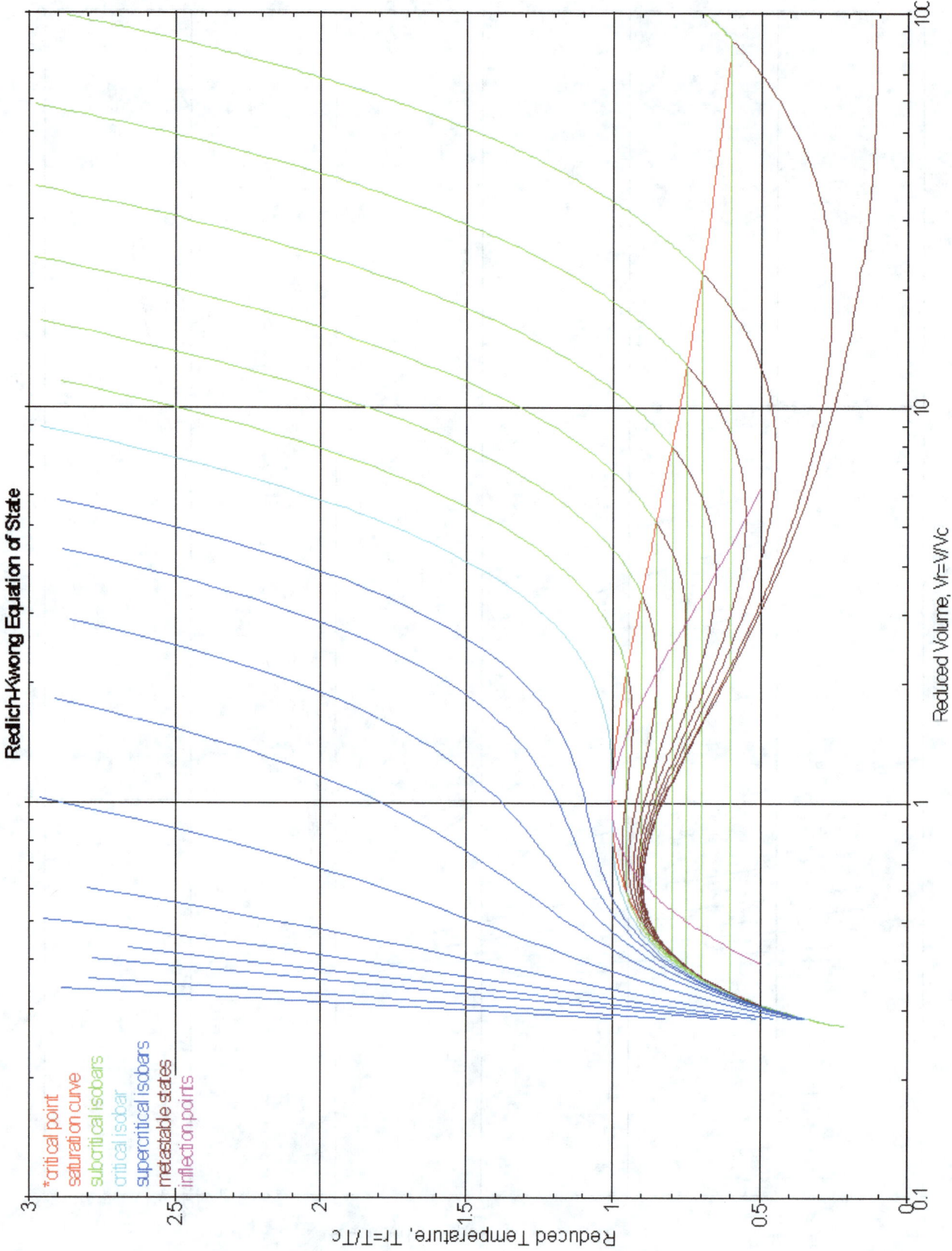

Figure 28. Tr vs. Vr Based on Redlich-Kwong

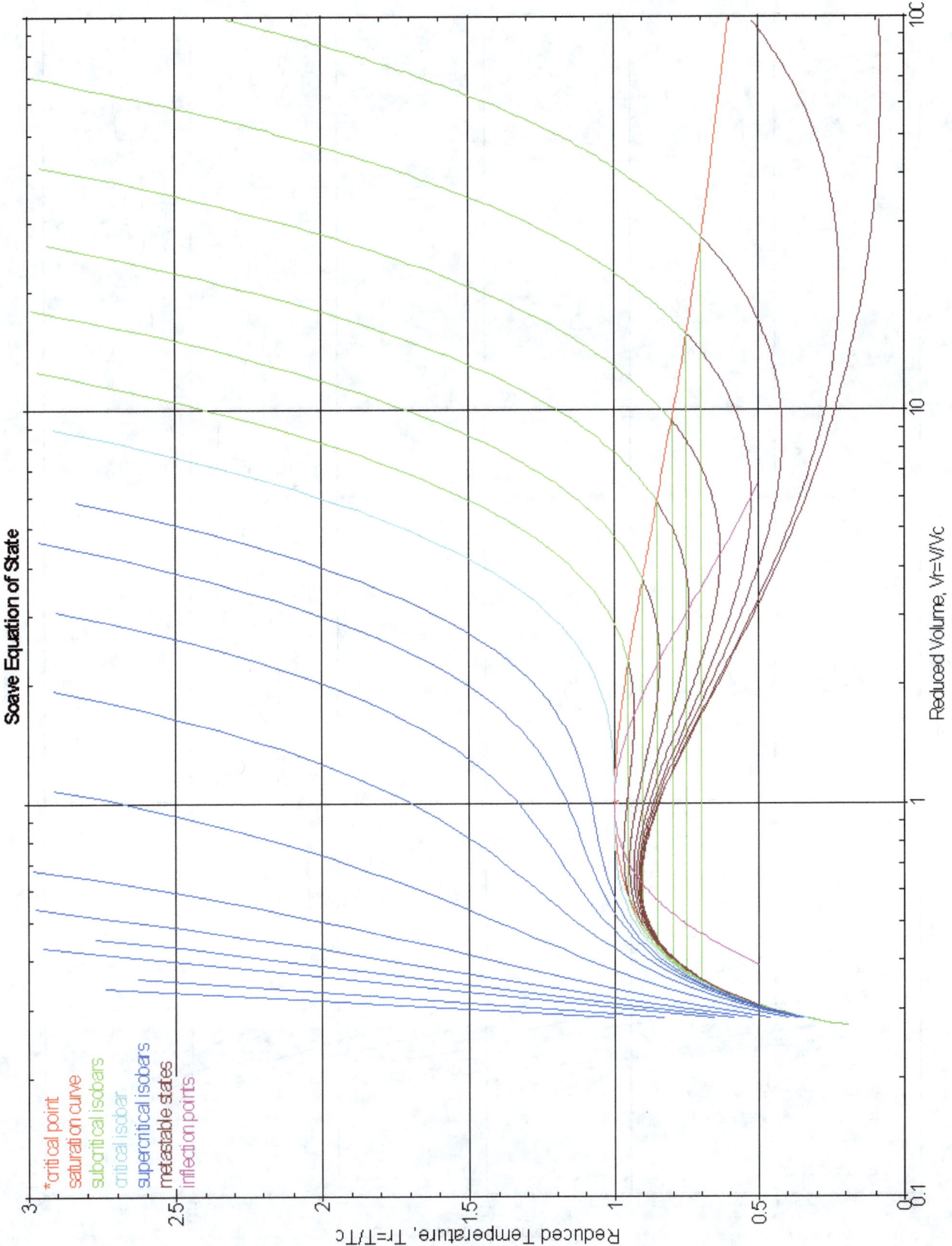

Figure 29. Tr vs. Vr Based on Soave's Modification

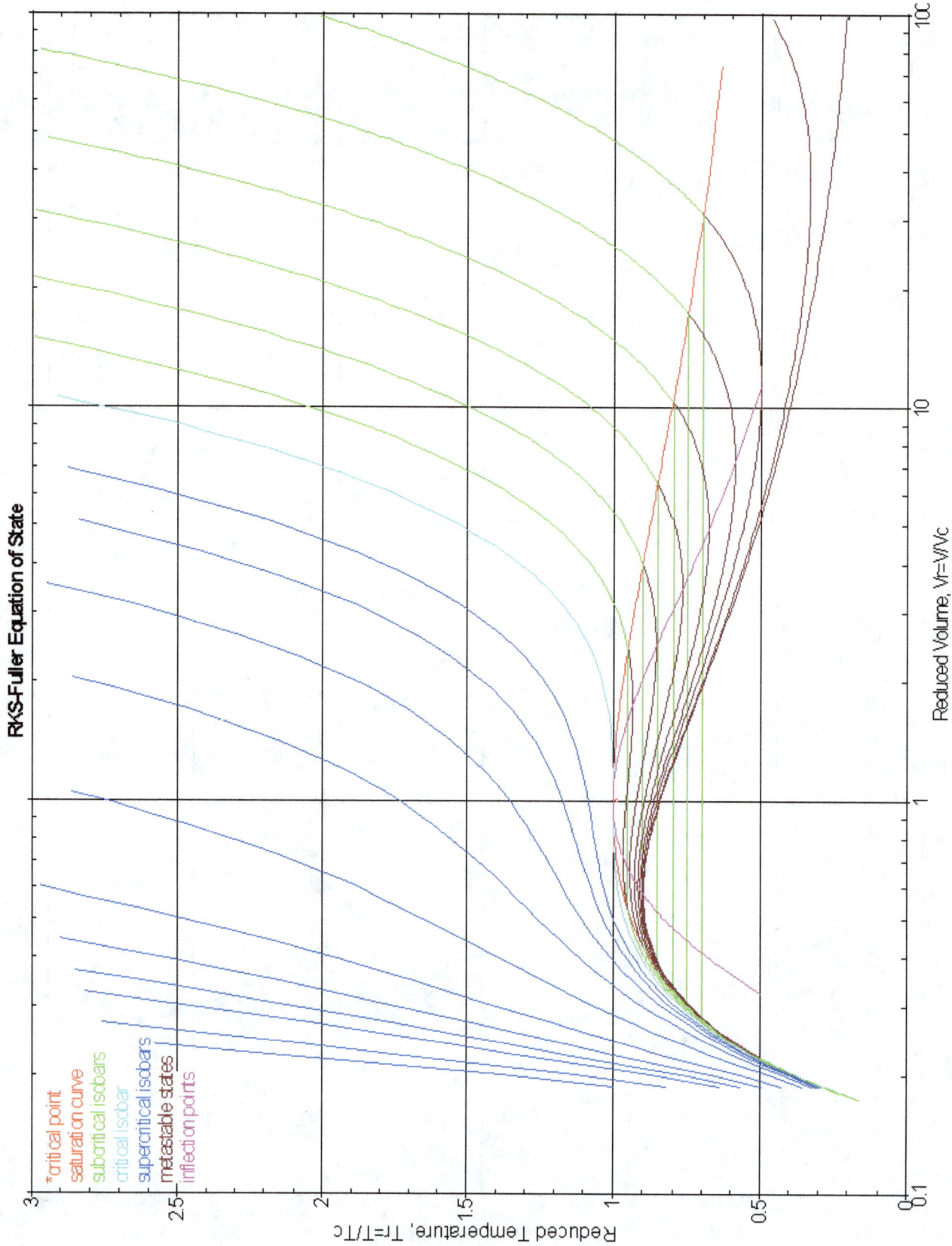

Figure 30. Tr vs. Vr Based on Fuller's Modification

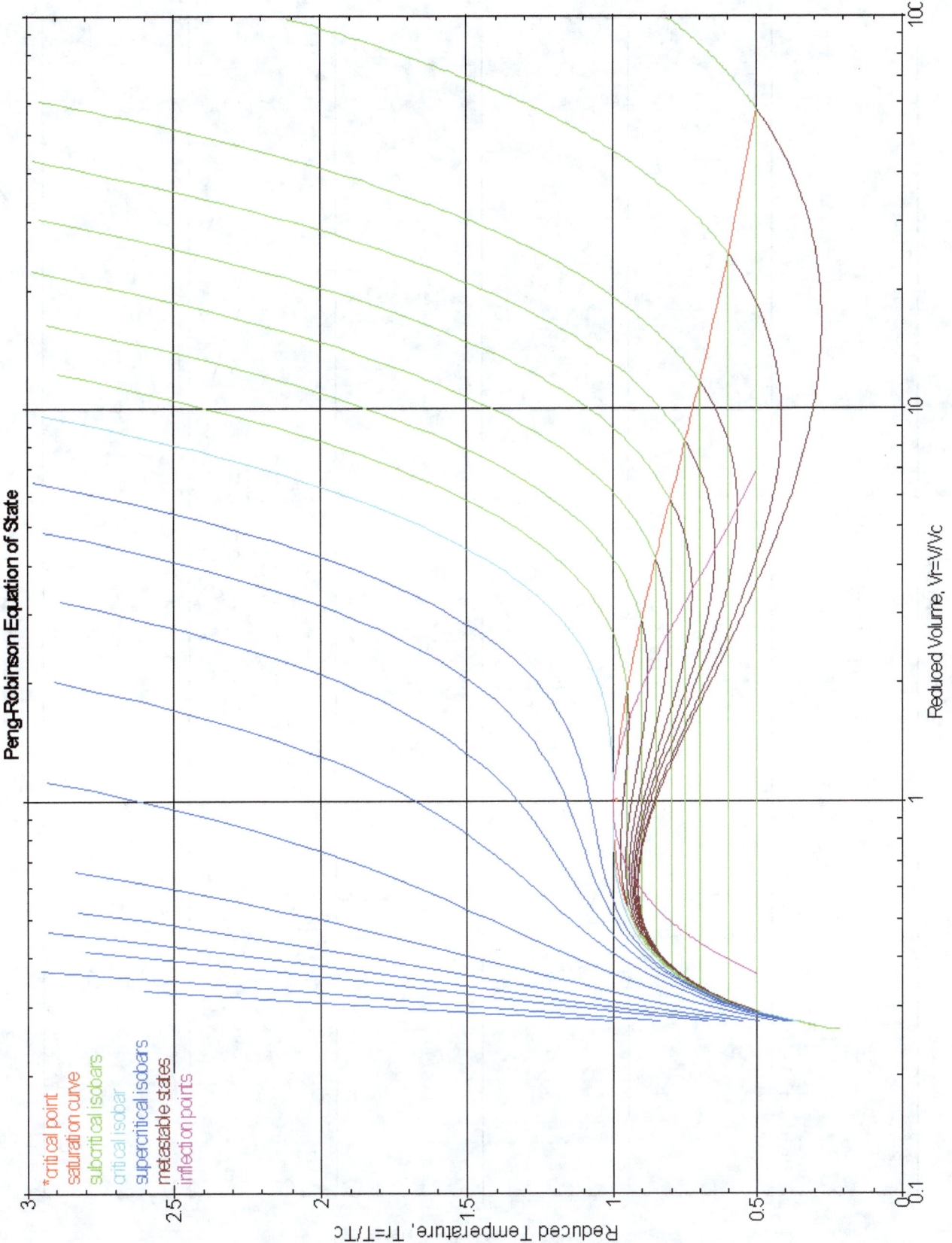

Figure 31. Tr vs. Vr Based on Peng-Robinson

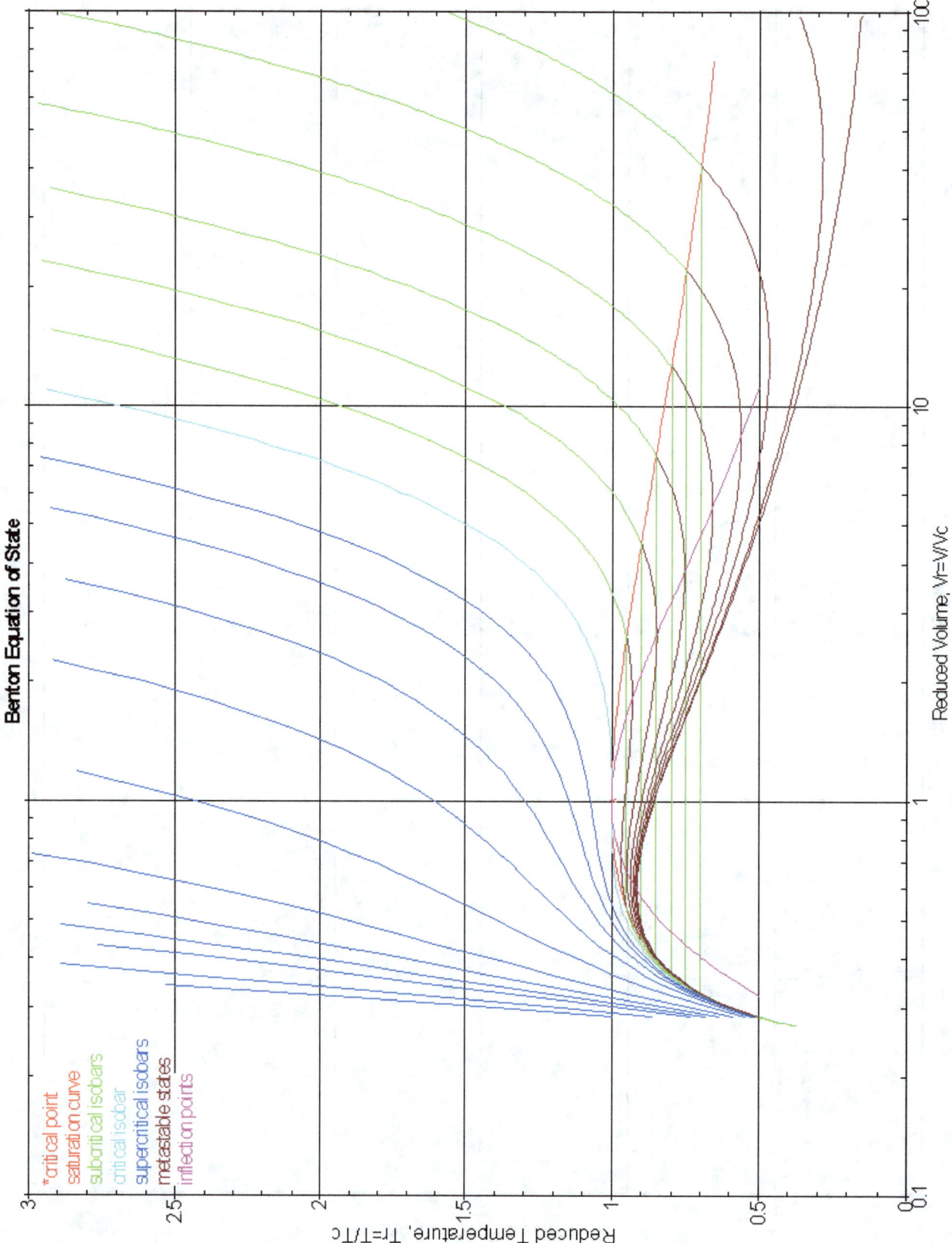

Figure 32. Tr vs. Vr Based on Author's Modification

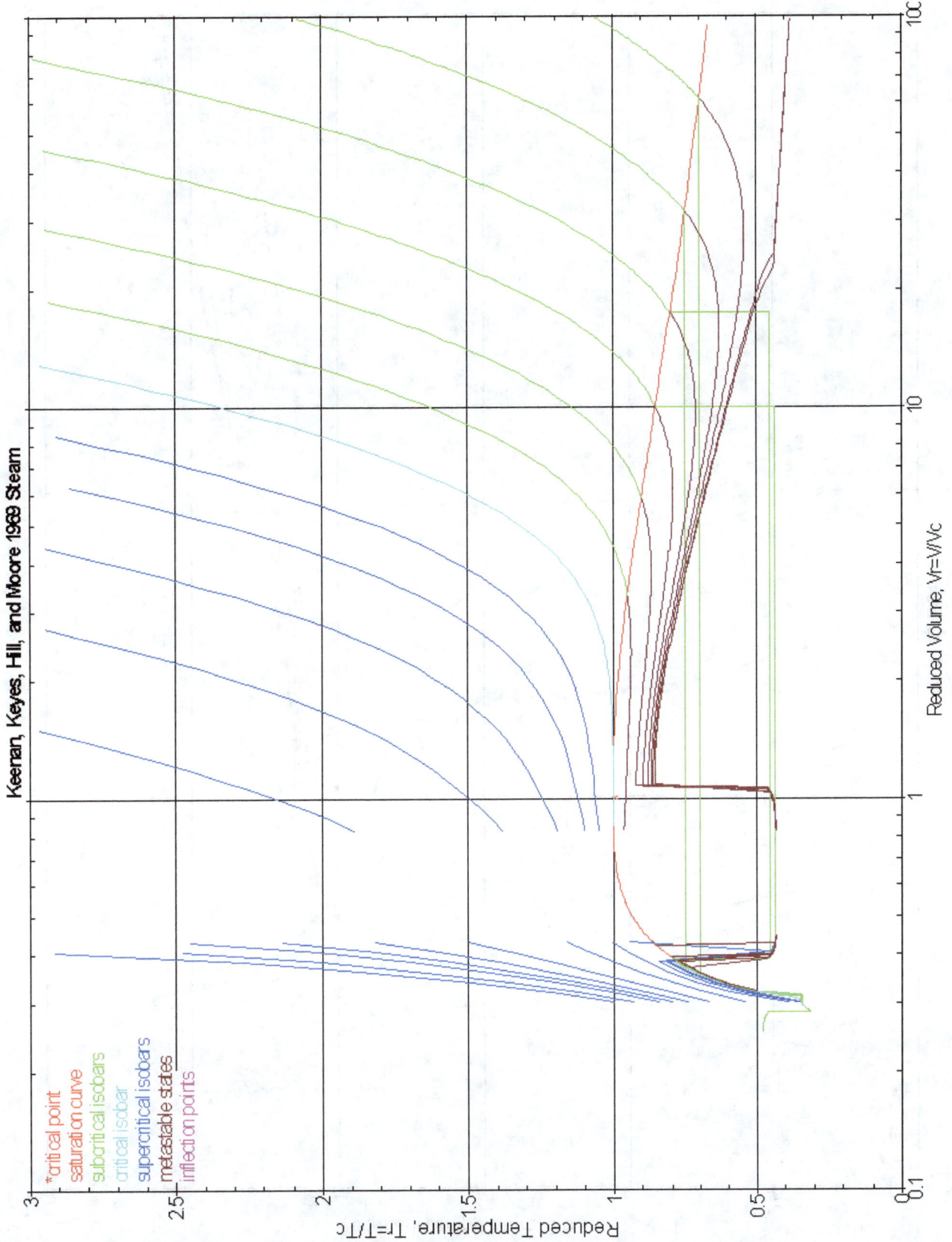

Figure 33. Tr vs. Vr Based on Keenan, Keyes, Hill, and Moore

The metastable region is undefined for this formulation.

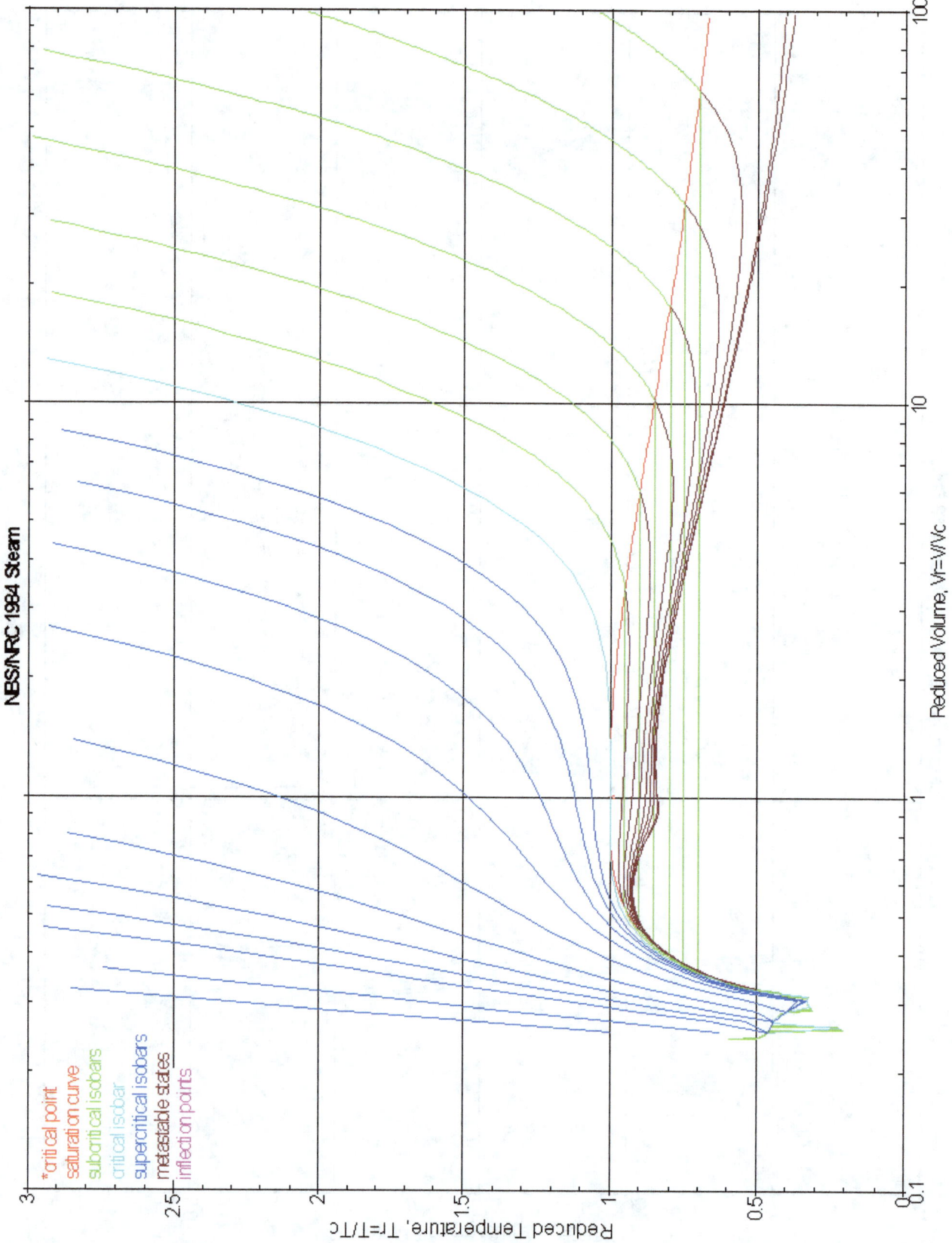

Figure 34. Tr vs. Vr Based on Haar, Gallagher, and Kell

The metastable region is undefined for this formulation.

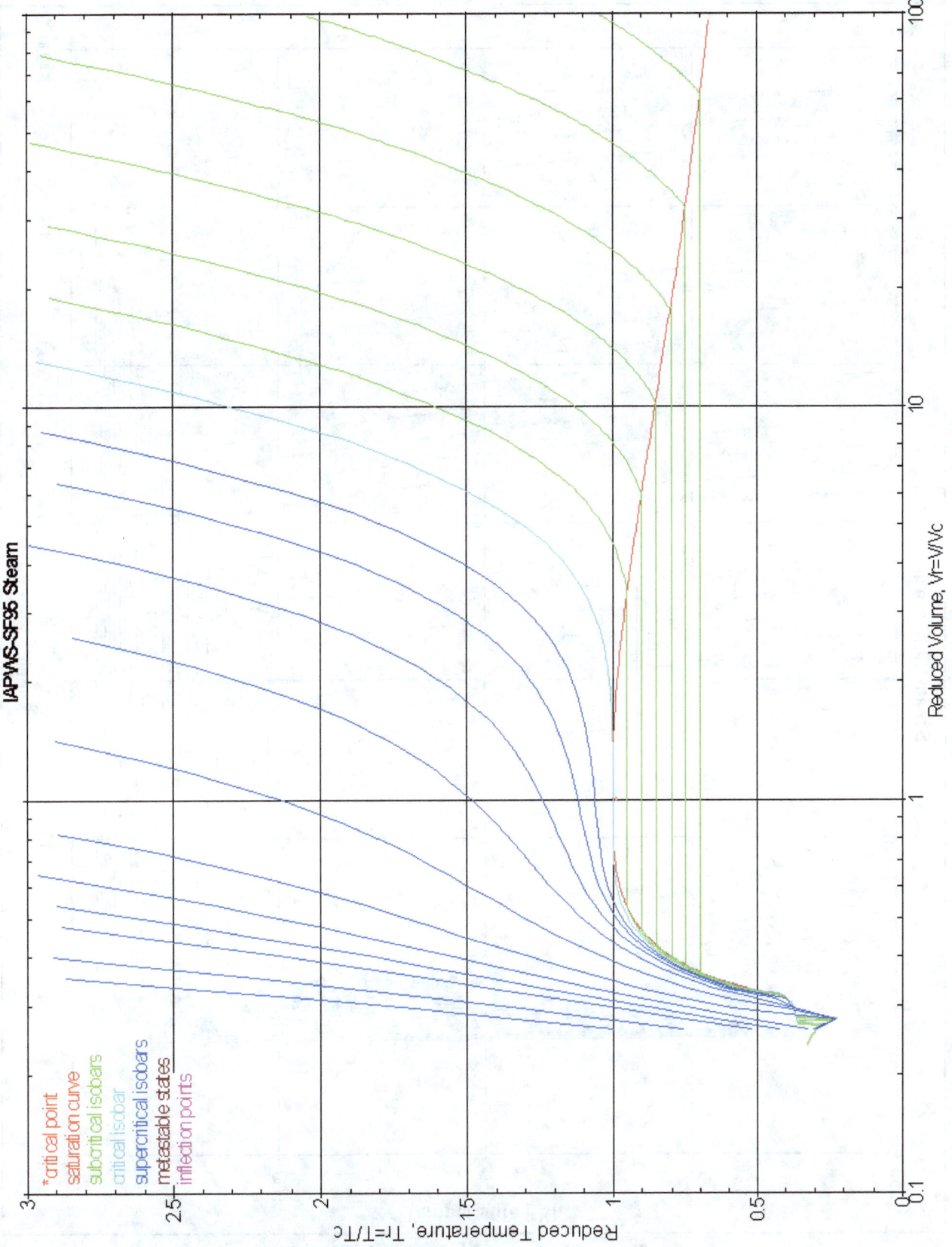

Figure 35. Tr vs. Vr Based on Wagner and Pruß

The metastable region is undefined for this formulation.

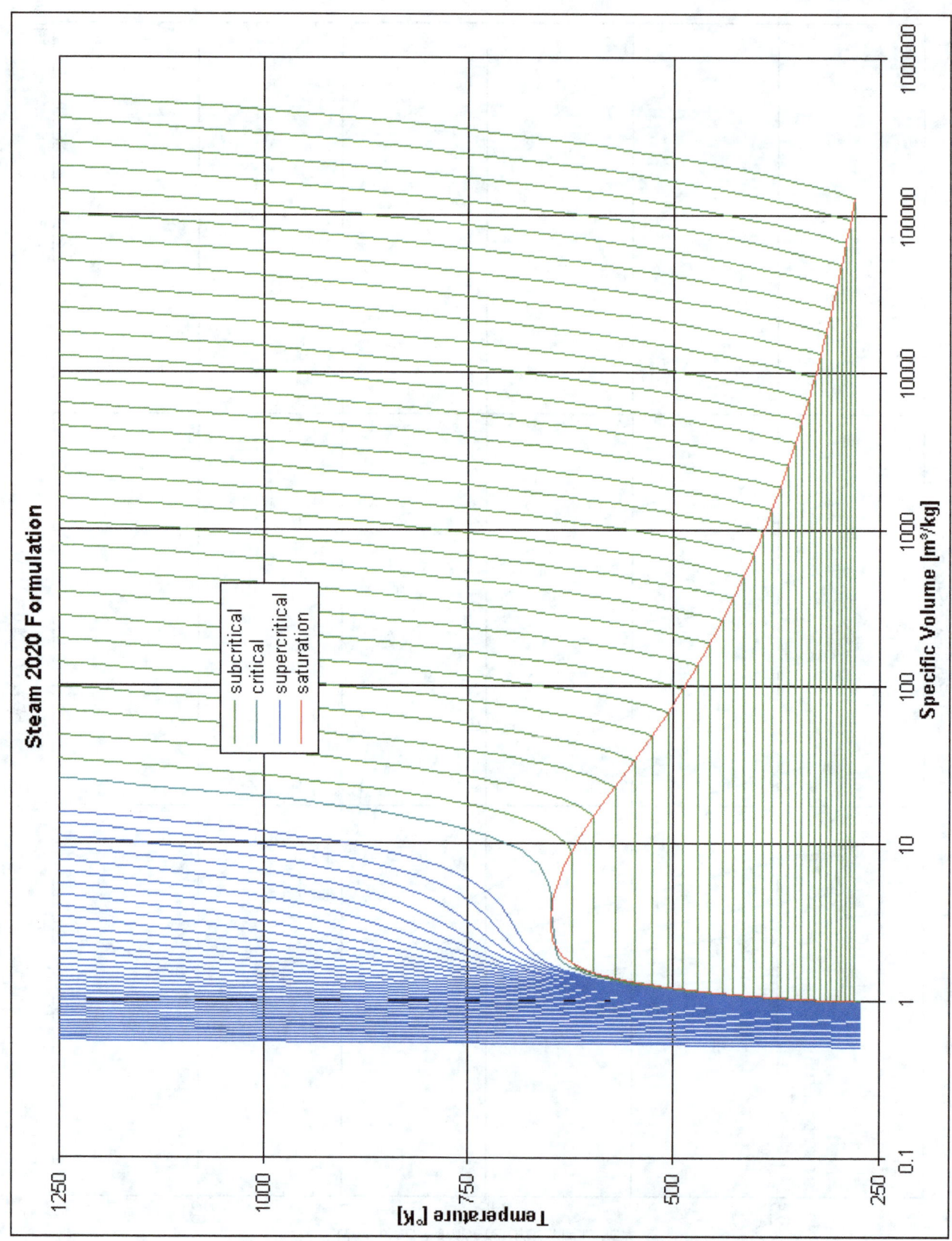

Figure 36. Tr vs. Vr Based on Steam 2020 Formulation

The metastable region is well-defined for this formulation but not shown in this figure.

Chapter 4. Compressibility

Compressibility, Z, was introduced with Equation 1.2 and then again for van der Waals with Equation 1.4. In this chapter we compare graphs of compressibility vs. pressure for constant values of temperature (isotherms). This is the original form of the Nelson-Obert data.

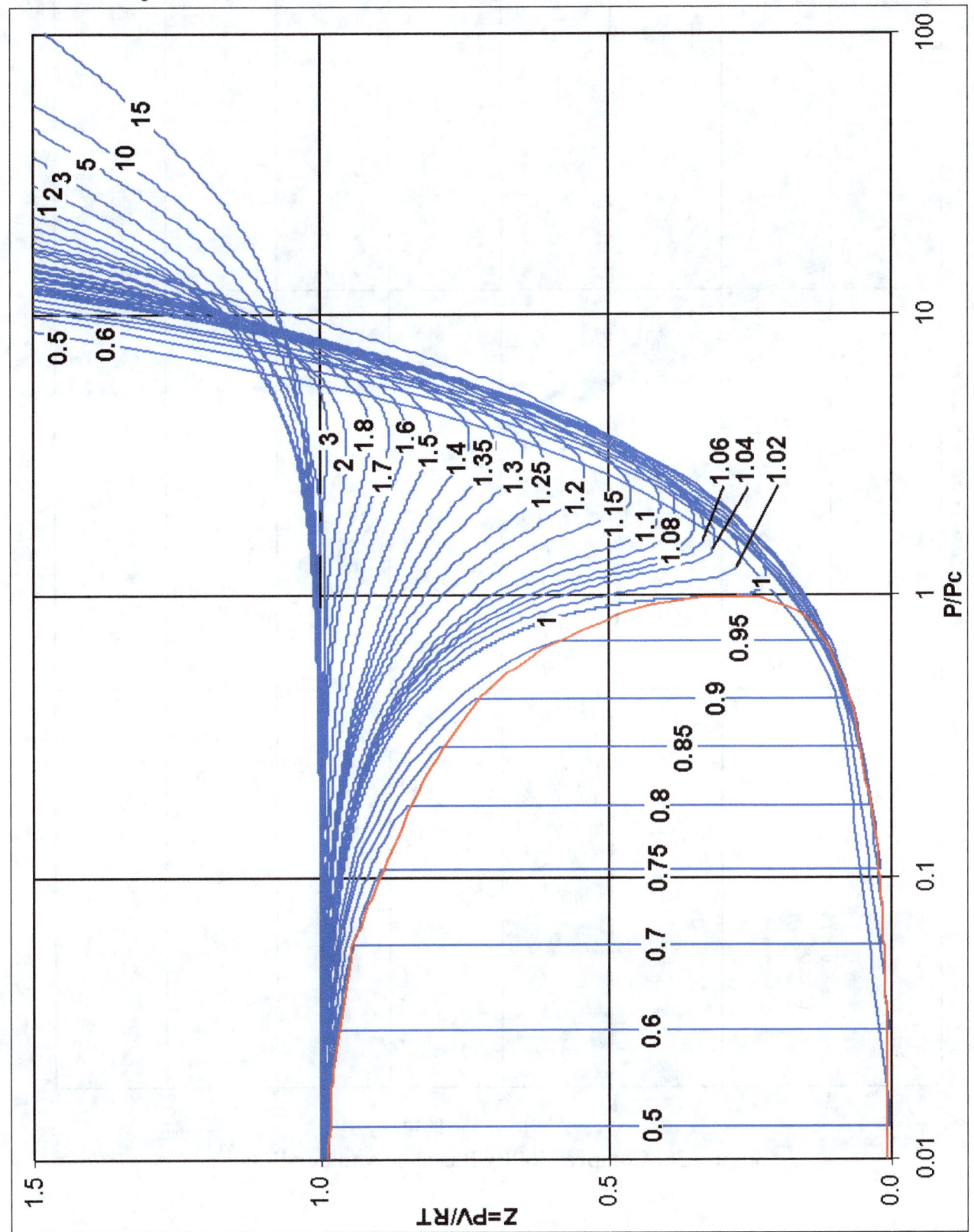

Figure 37. Compressibility Based on Nelson-Obert

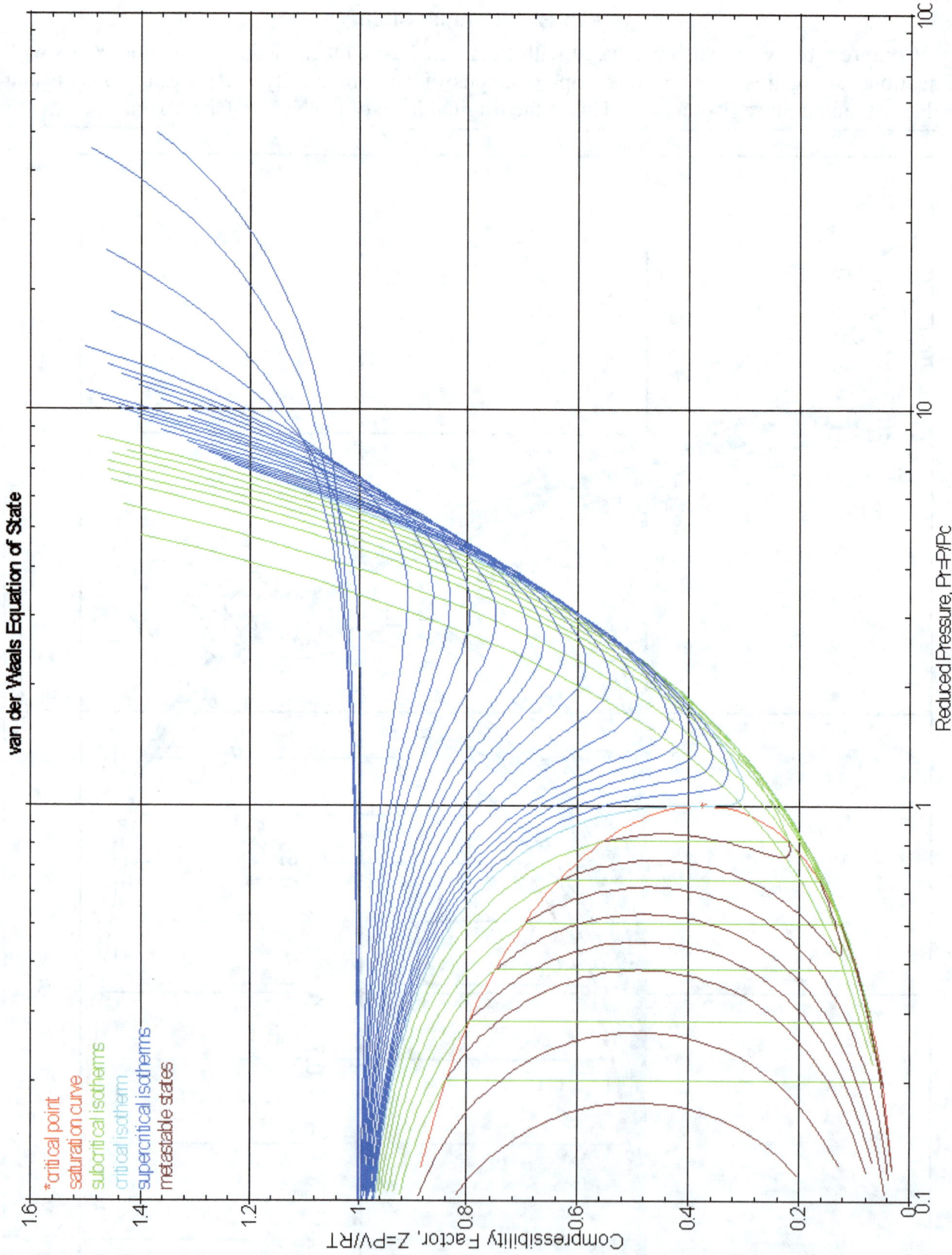

Figure 38. Compressibility Based on van der Waals

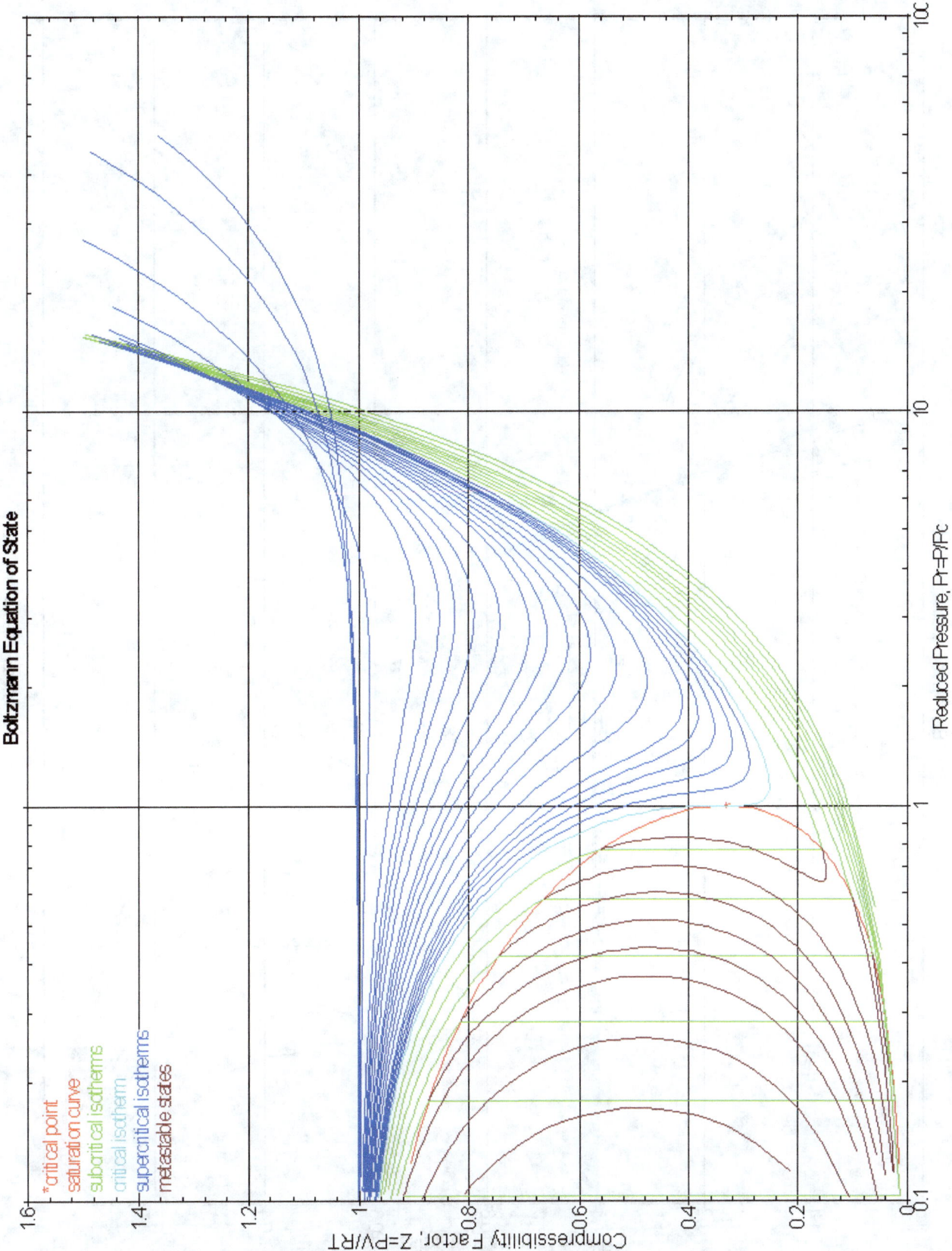

Figure 39. Compressibility Based on Boltzmann

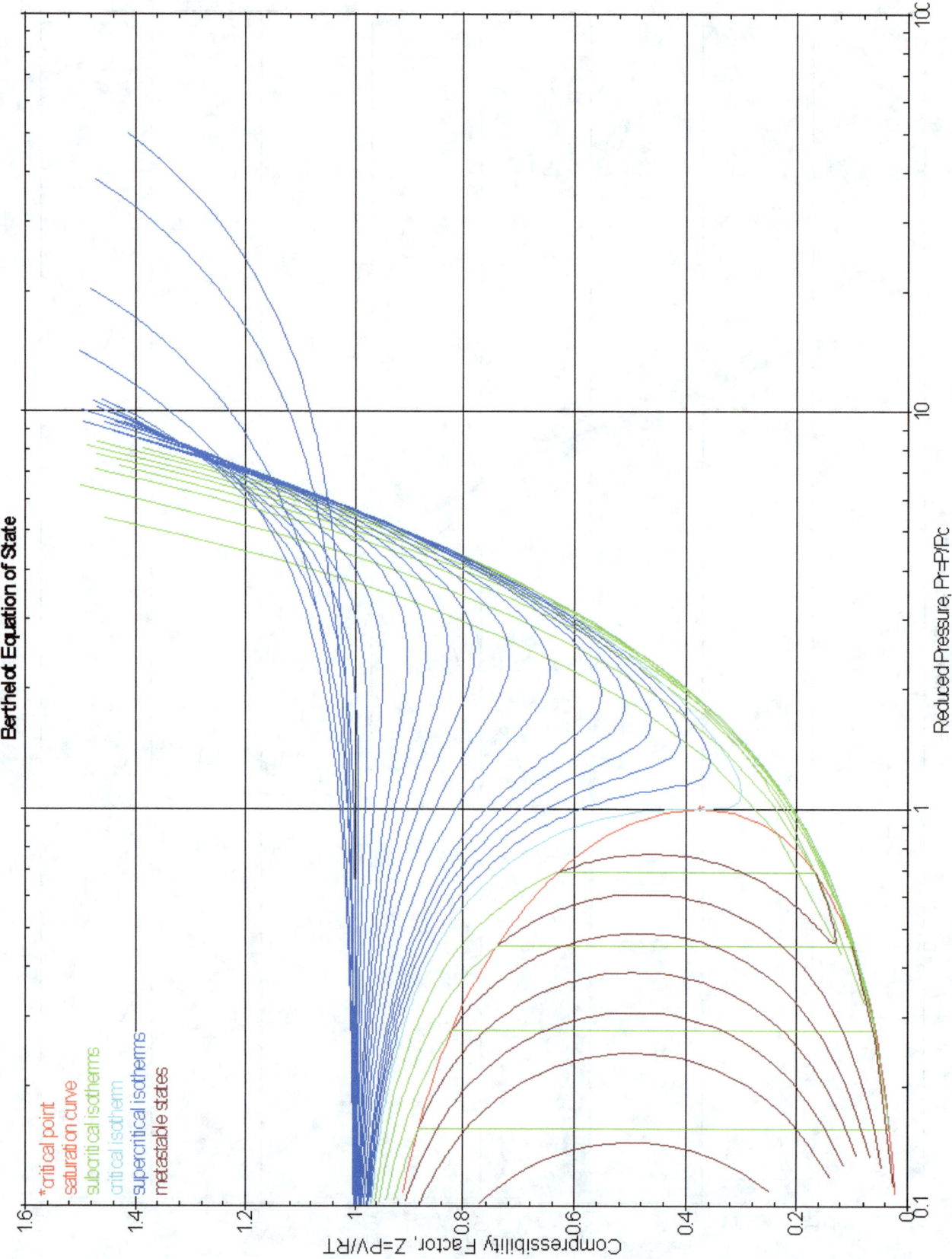

Figure 40. Compressibility Based on Berthelot

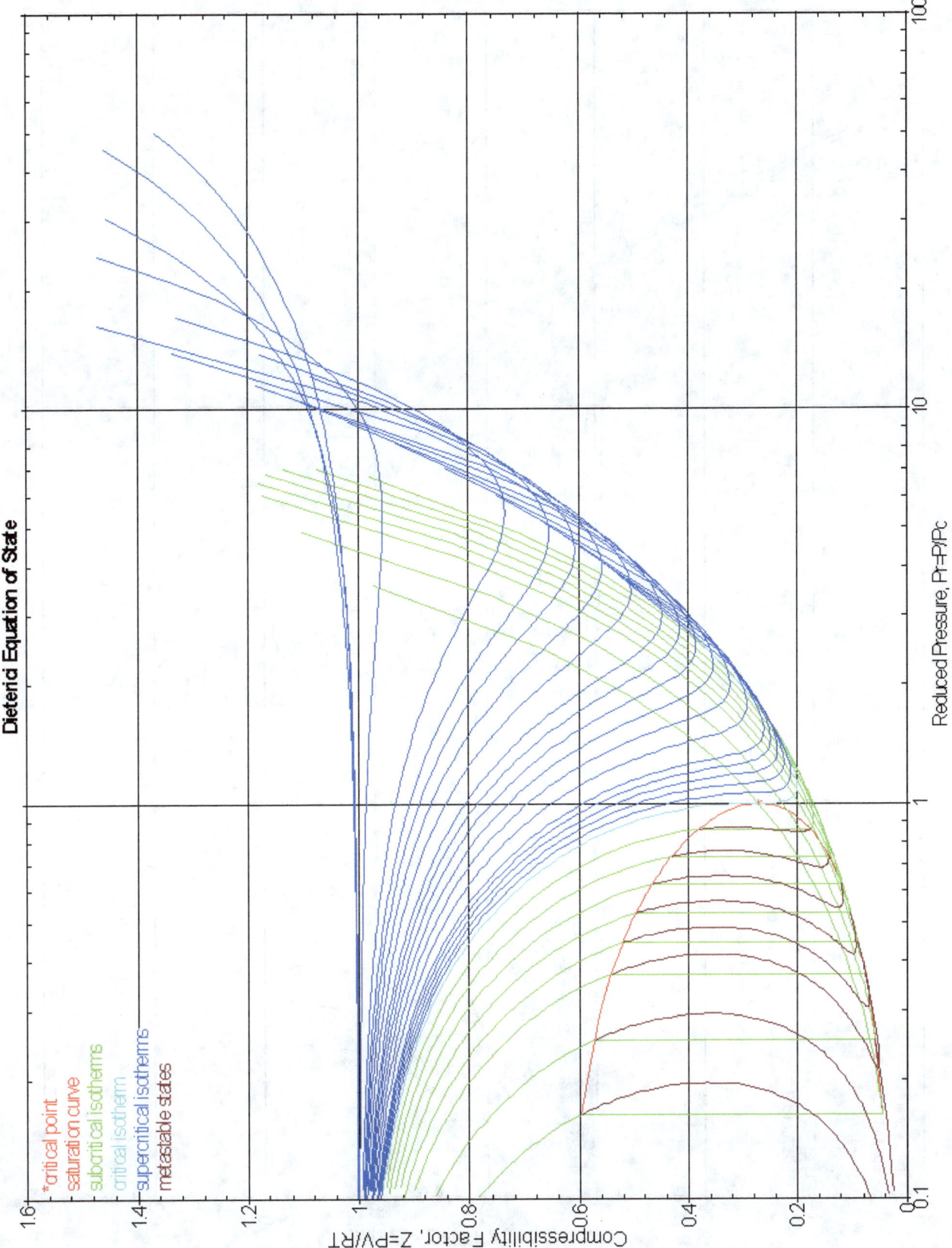

Figure 41. Compressibility Based on Dieterici

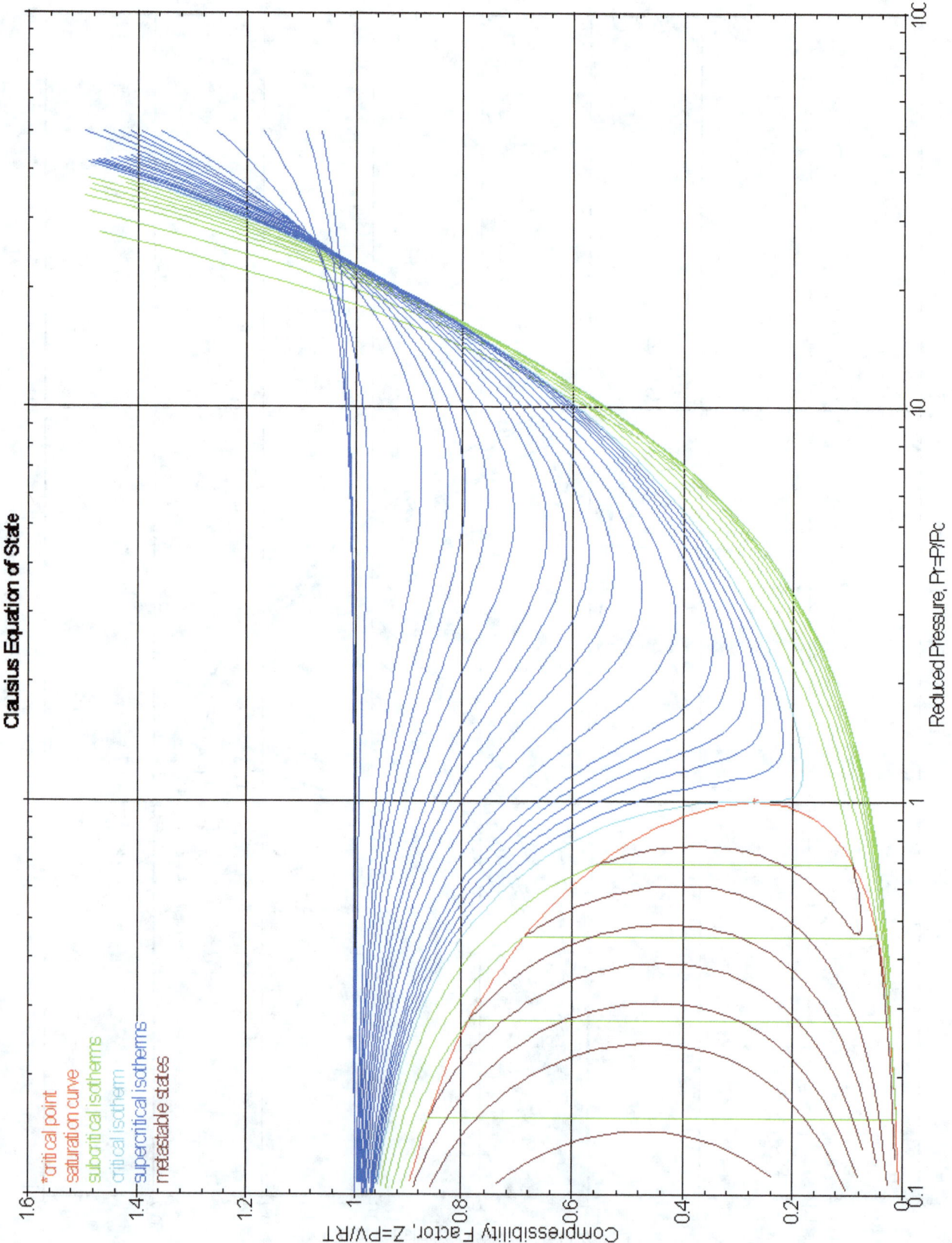

Figure 42. Compressibility Based on Clausius

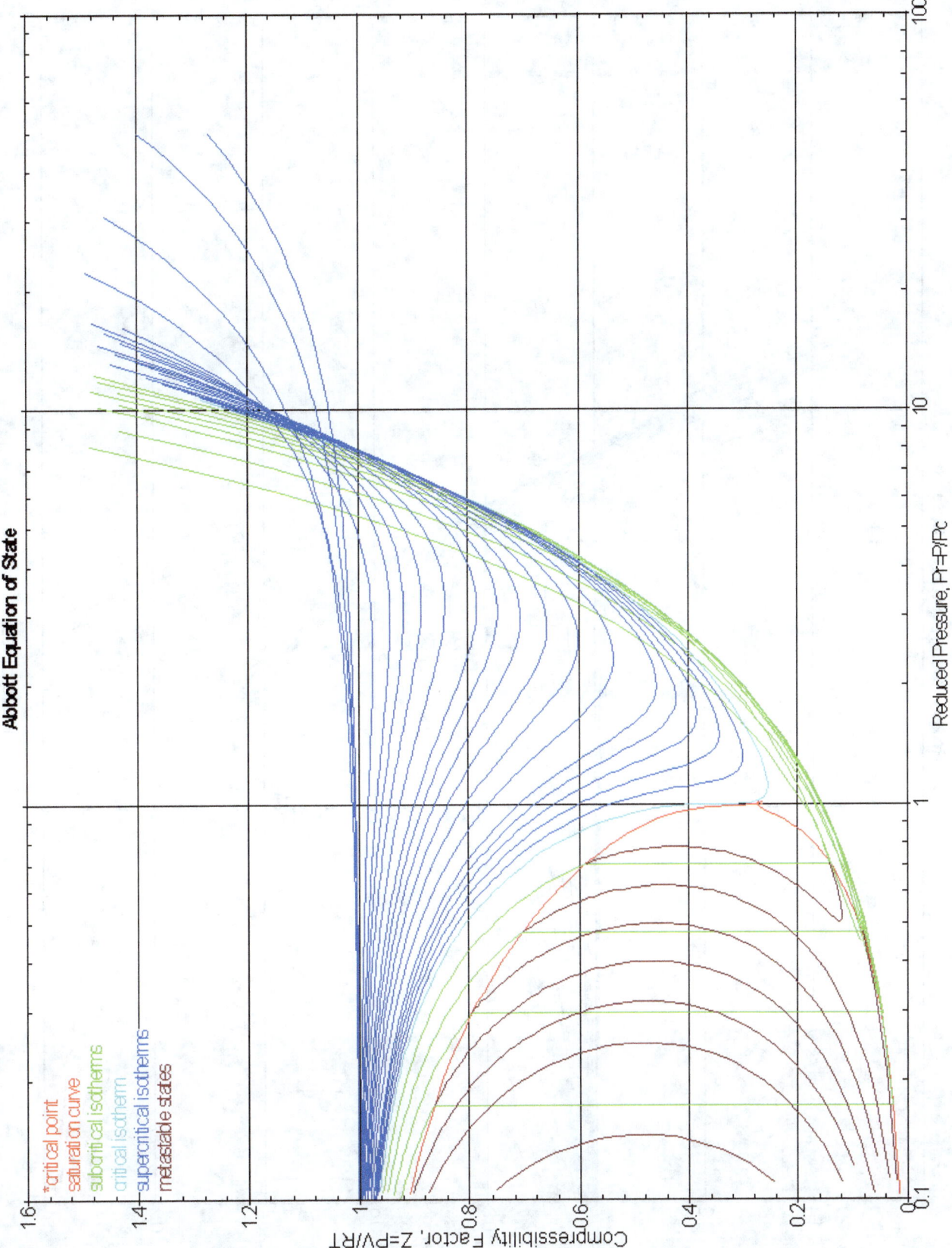

Figure 43. Compressibility Based on Abbott's Modification

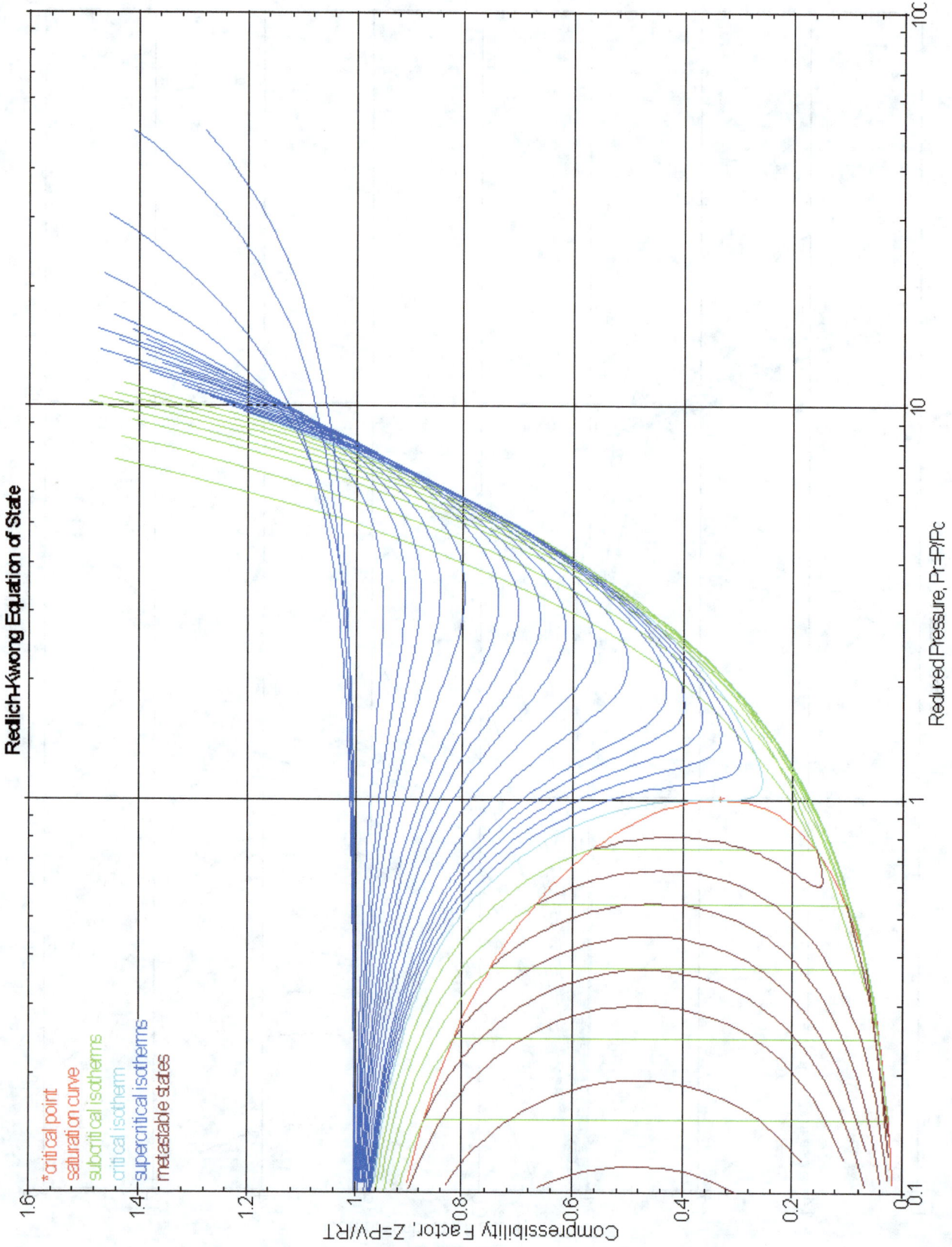

Figure 44. Compressibility Based on Redlich-Kwong

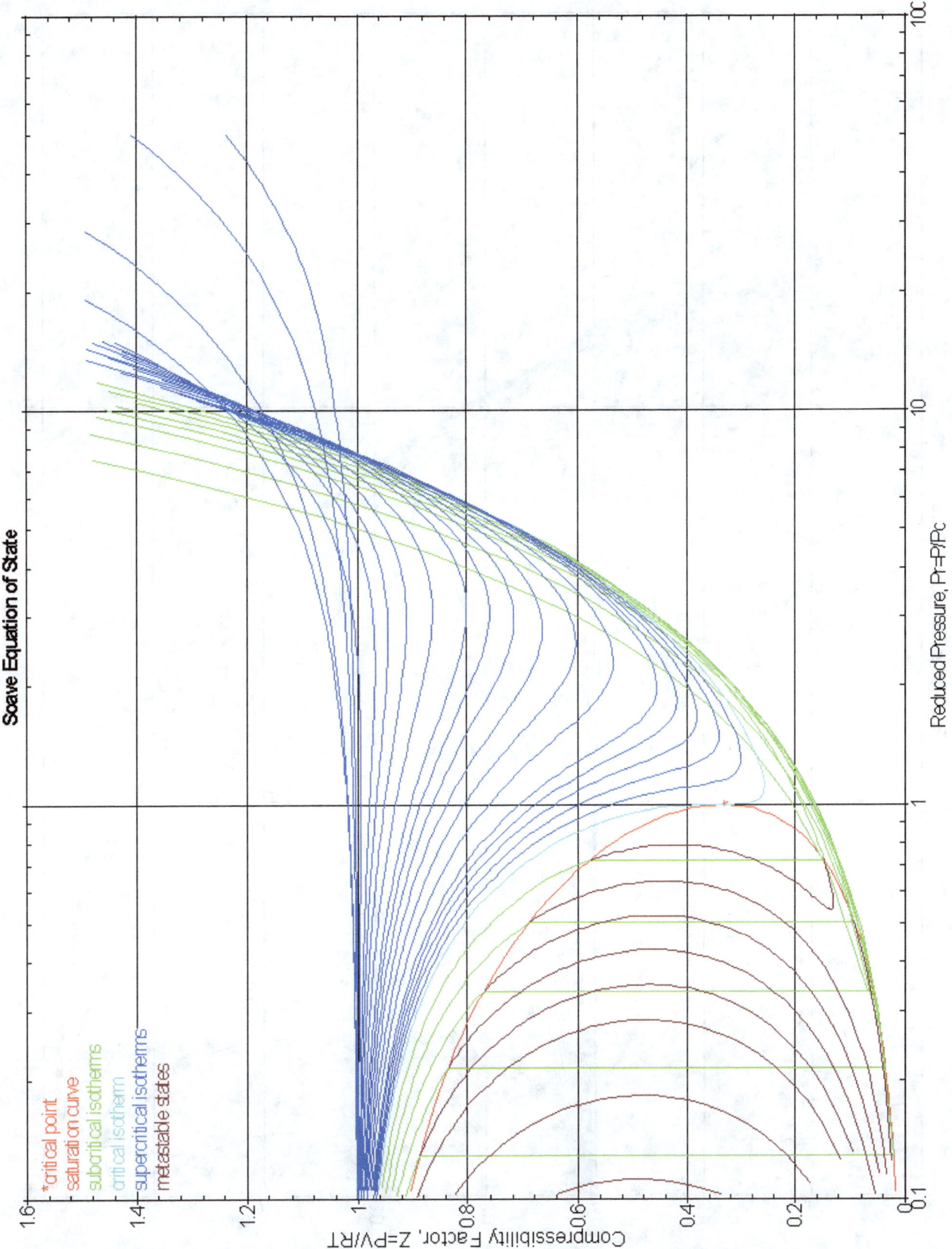

Figure 45. Compressibility Based on Soave's Modification

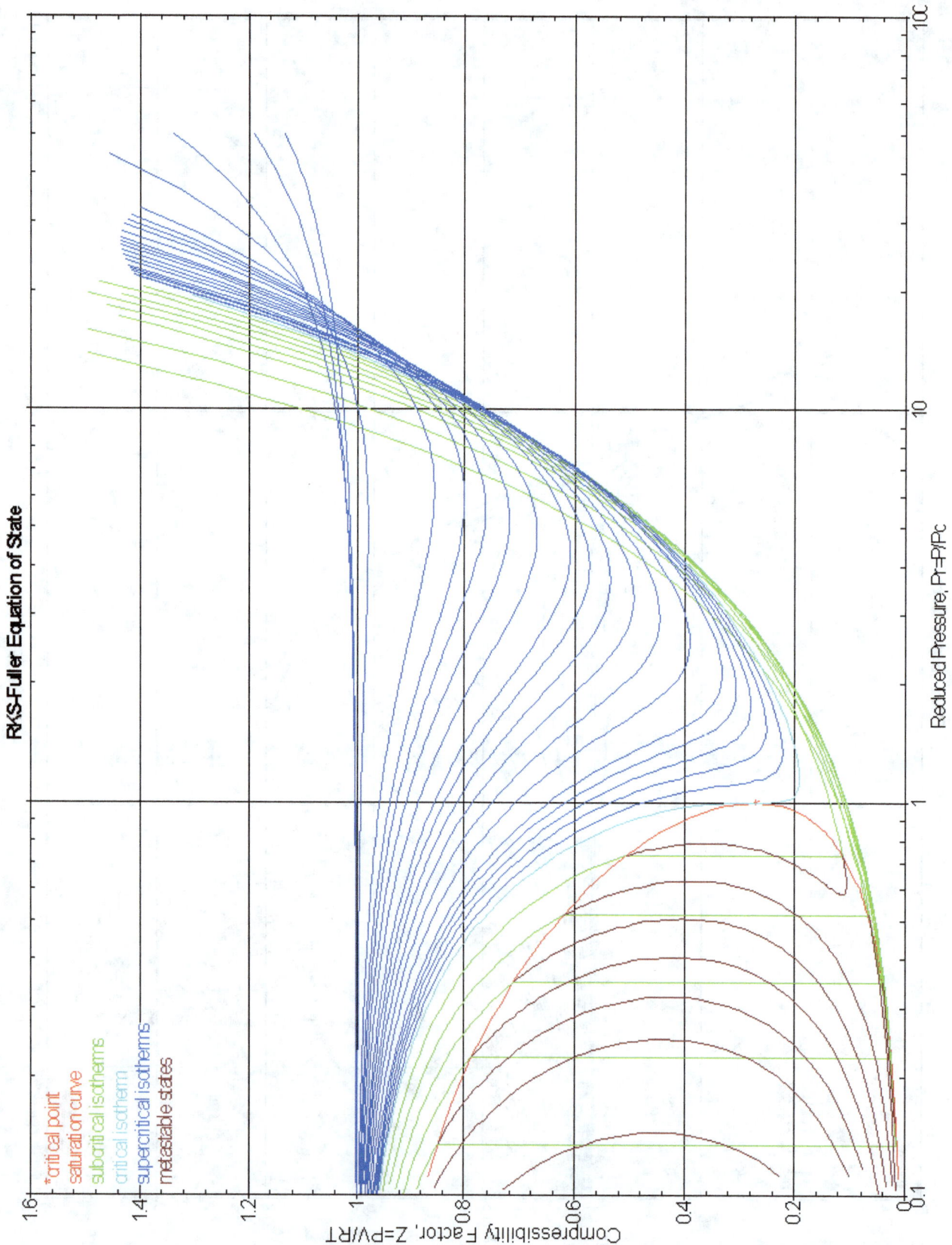

Figure 46. Compressibility Based on Fuller's Modification

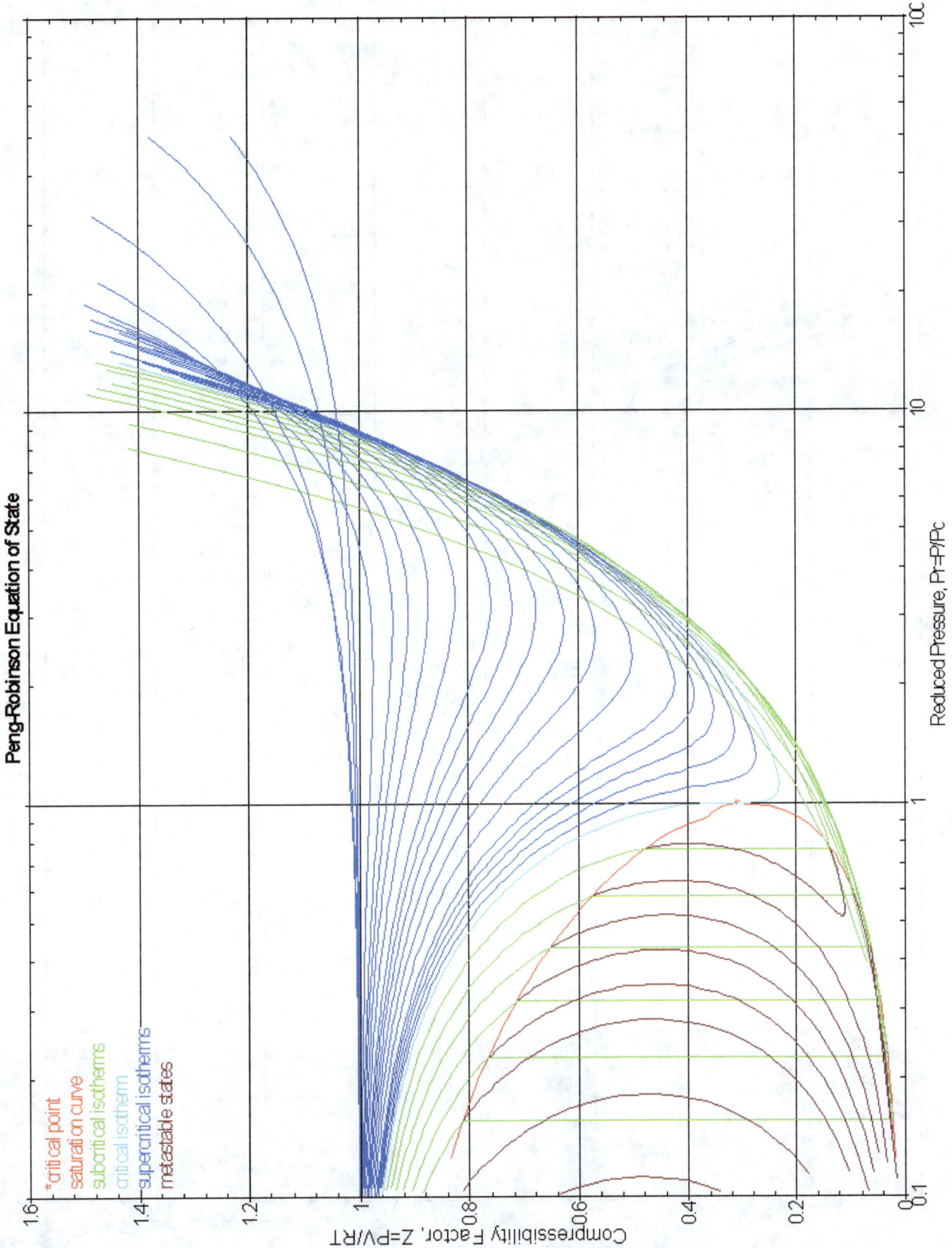

Figure 47. Compressibility Based on Peng-Robinson

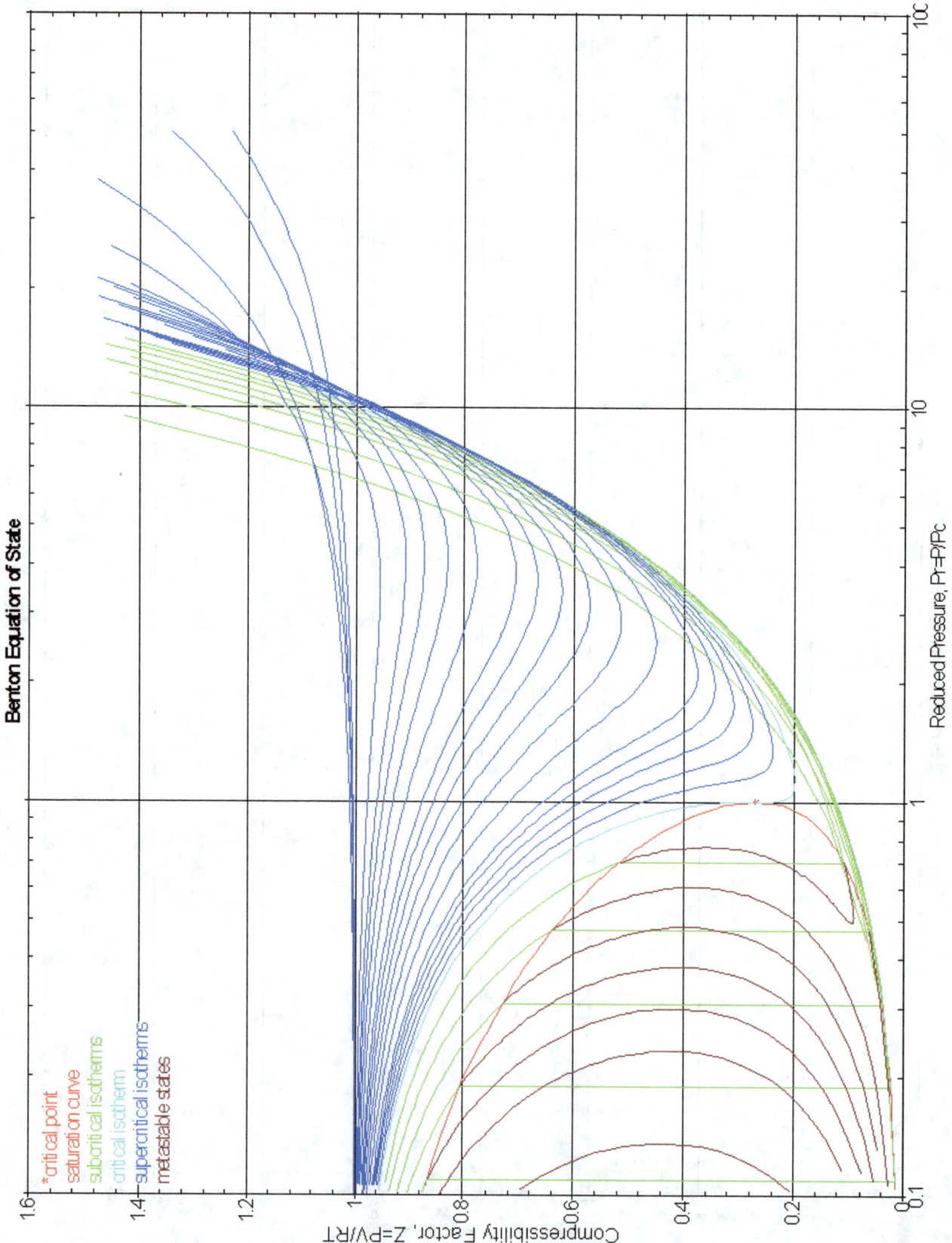

Figure 48. Compressibility Based on Author's Modification

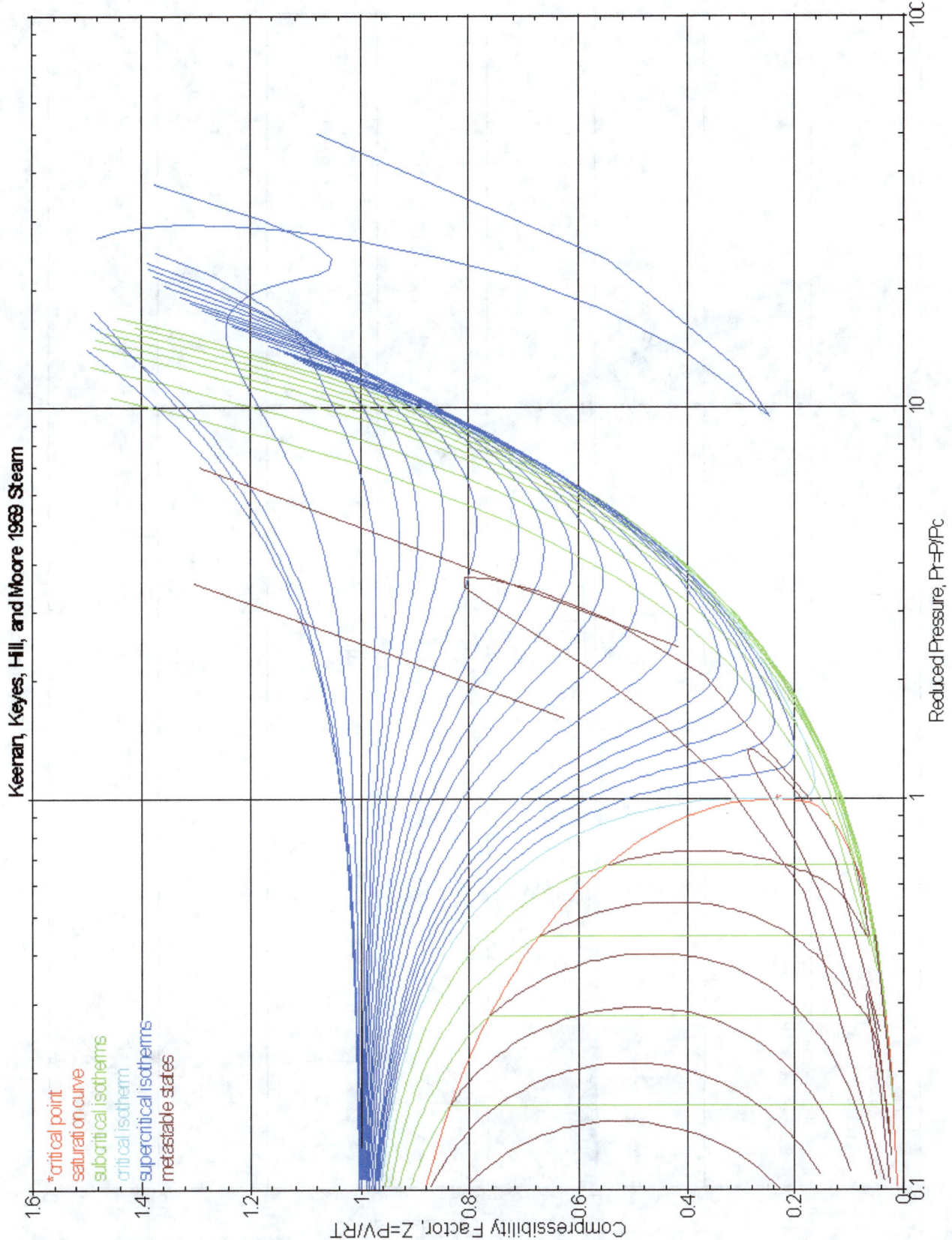

Figure 49. Compressibility Based on Keenan, Keyes, Hill, and Moore

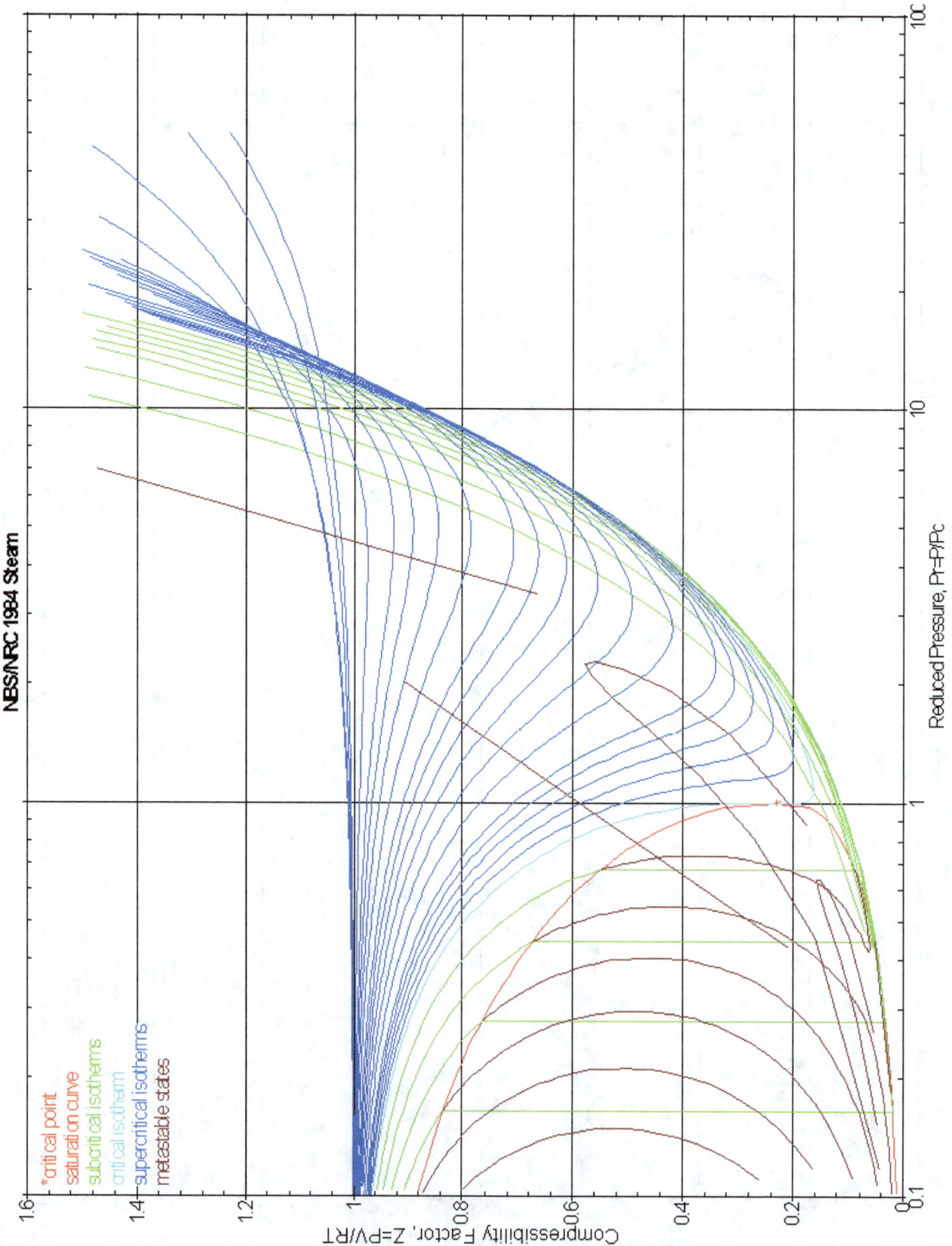

Figure 50. Compressibility Based on Haar, Gallagher, and Kell

The metastable region is undefined for this formulation.

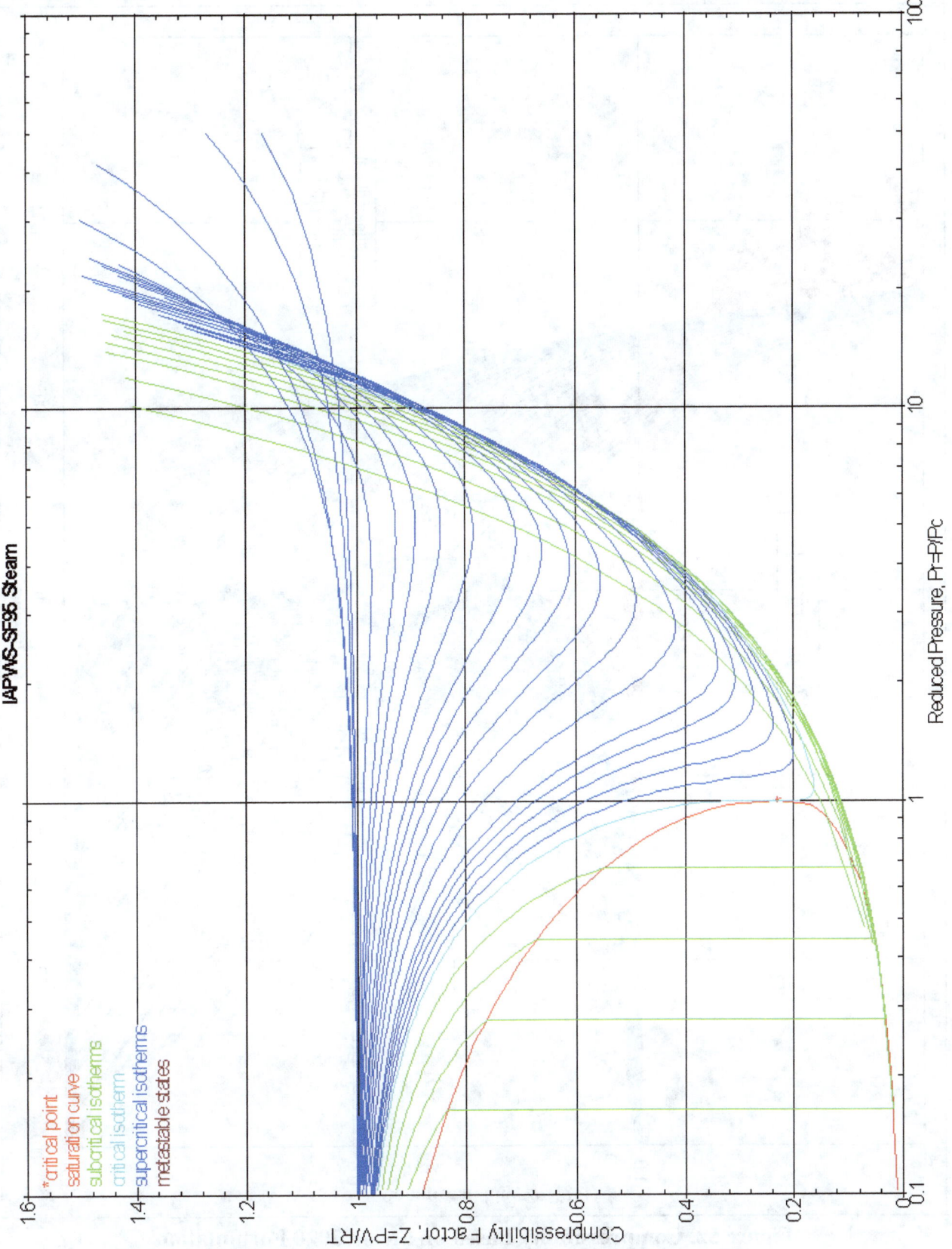

Figure 51. Compressibility Based on Wagner and Pruß

The metastable region is undefined for this formulation.

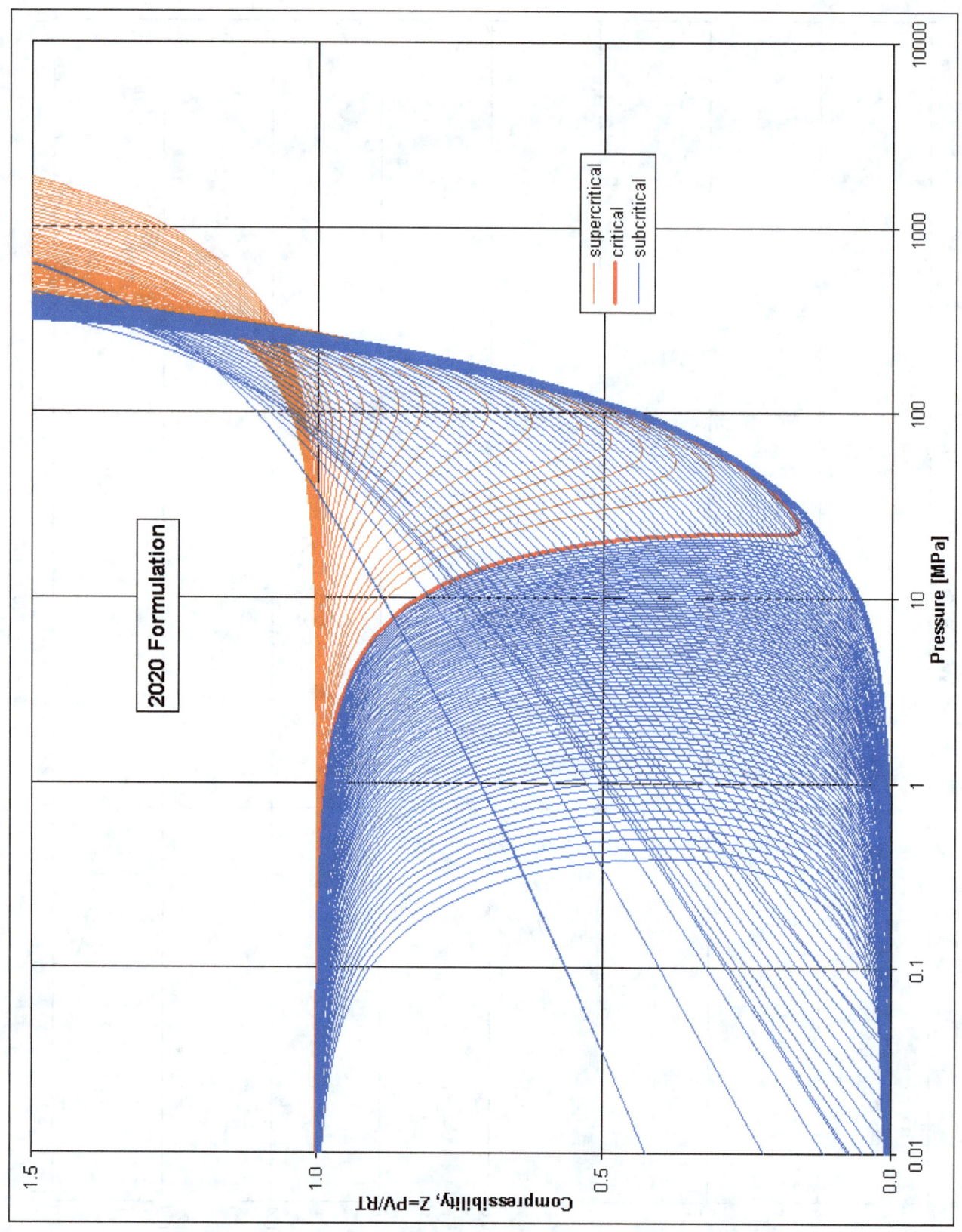

Figure 52. Compressibility Based on Steam 2020 Formulation

Chapter 5. Fugacity

Fugacity (given the symbol, F) is a sort of pseudo-pressure. It is often defined as the pressure that would result in an ideal gas having the same Gibbs free energy *(g=h-Ts)* as the real gas (at the same temperature). We don't ever use fugacity in this way, but it does provide a conceptual framework. Most often, we consider the fugacity coefficient (given the symbol, φ), which is the ratio of fugacity to pressure, making it dimensionless. There are several derivations and forms used for the fugacity. The derivations are beyond the scope of this work. For our purposes, we will use the following:

$$\ln \varphi = \int_0^P (Z-1) \frac{dP}{P} \tag{5.1}$$

Equation 5.1 is rarely practical to evaluate. Integration by parts yields the following much more useful formula:

$$\ln \varphi = Z - 1 - \ln Z - \int_\infty^V \left(\frac{P}{RT} - \frac{1}{V} \right) dV \tag{5.2}$$

Some of the equations of state used in this text can be analytically integrated to yield a reasonably compact expression; for example, the van der Waals:

$$\ln \varphi = Z - 1 - \ln Z + \ln\left(\frac{V_R}{V_R - B} \right) - \frac{A}{V_R T_R} \tag{5.3}$$

Fugacity coefficient for the Berthelot can be expressed as.

$$\ln \varphi = Z - 1 - \ln Z + \ln\left(\frac{V_R}{V_R - B} \right) - \frac{A}{V_R T_R^2} \tag{5.4}$$

The Clausius equivalent is:

$$\ln \varphi = Z - 1 - \ln Z + \ln\left(\frac{V_R}{V_R - B} \right) - \frac{A}{(V_R + C) T_R^2} \tag{5.5}$$

Abbott's modification yields:

$$\ln \varphi = Z - 1 - \ln Z + \ln\left(\frac{V_R}{V_R - B} \right) - \frac{A}{(V_R + C) T_R} \tag{5.6}$$

The Redlich-Kwong fugacity coefficient is (Soave's modification is nearly the same):

$$\ln \varphi = Z - 1 - \ln Z + \ln\left(\frac{V_R}{V_R - B} \right) - \frac{A}{B T_R} \ln\left(\frac{V_R + B}{V_R} \right) \tag{5.7}$$

The fugacity coefficient for Fuller's modification is:

$$\ln \varphi = Z - 1 - \ln Z + \ln\left(\frac{V_R}{V_R - B} \right) - \frac{A}{C T_R} \ln\left(\frac{V_R + C}{V_R} \right) \tag{5.8}$$

Calculation of the fugacity coefficient for the Dieterici EOS requires numerical integration. Gauss quadrature works well for this. The Peng-Robinson and this Author's modification to the cubic EOS have analytical solutions, but are quite lengthy and so not listed here. These can be found online. Fugacity coefficients for the steam formulations are also quite involved but manageable. Note that the fugacity is the same for saturated liquid and vapor so that there is no vapor dome in these next figures.

We begin with the Nelson-Obert data. Note that isotherms should never cross below about 5x the critical pressure on a fugacity graph. Note that all of the formulations shown below satisfy this requirement.

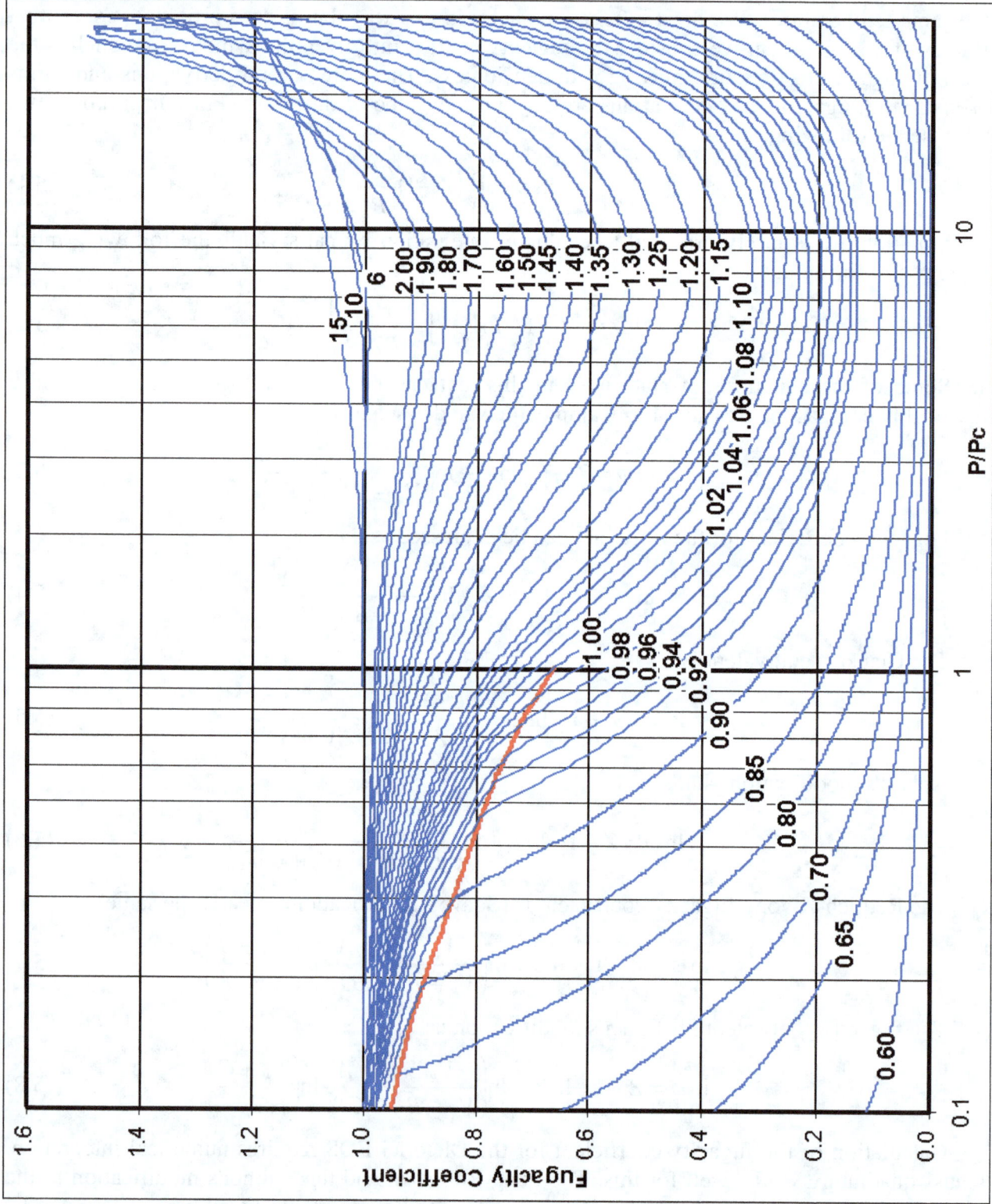

Figure 53. Fugacity Coefficient Based on the Nelson-Obert Generalized Data

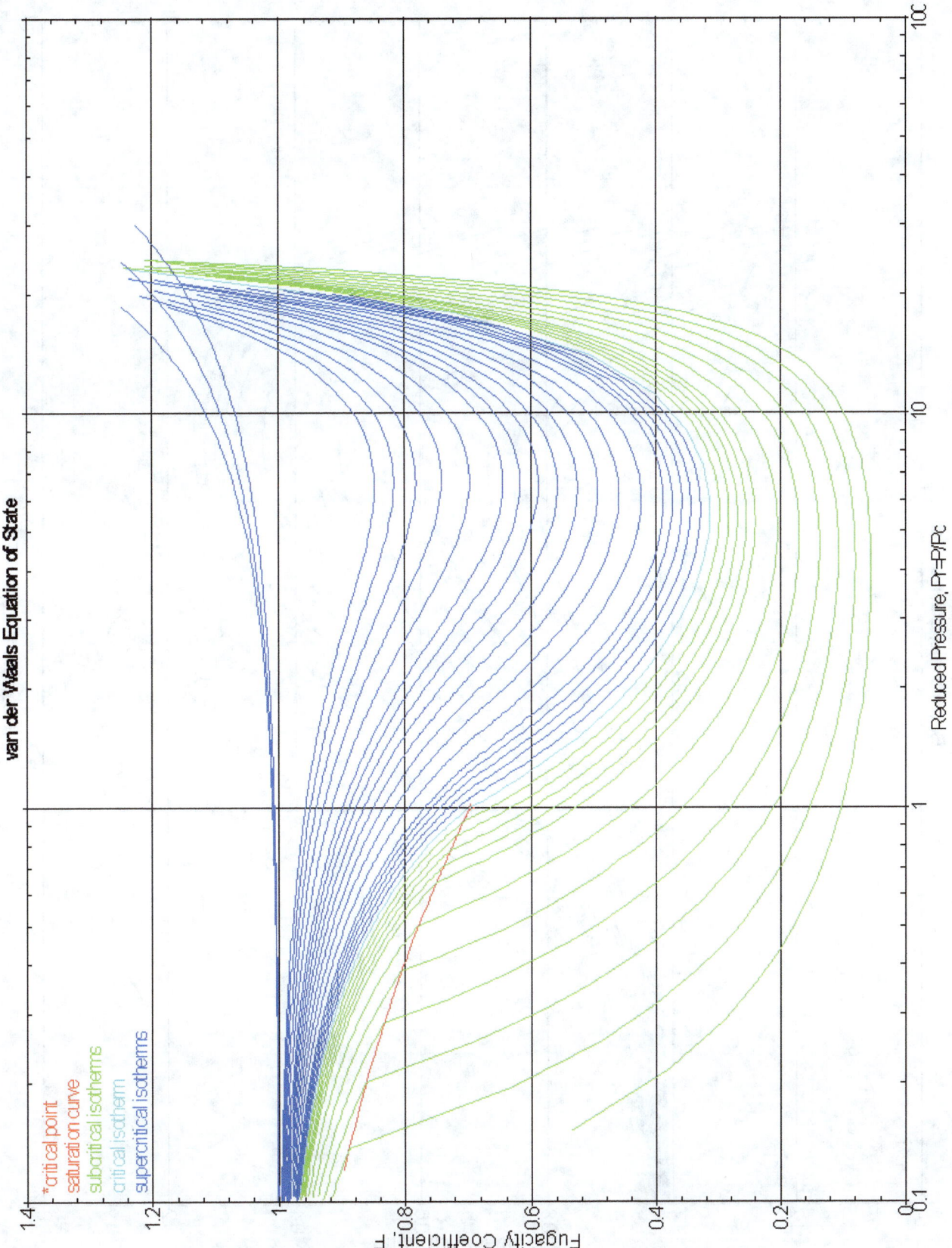

Figure 54. Fugacity Coefficient Based on van der Waals

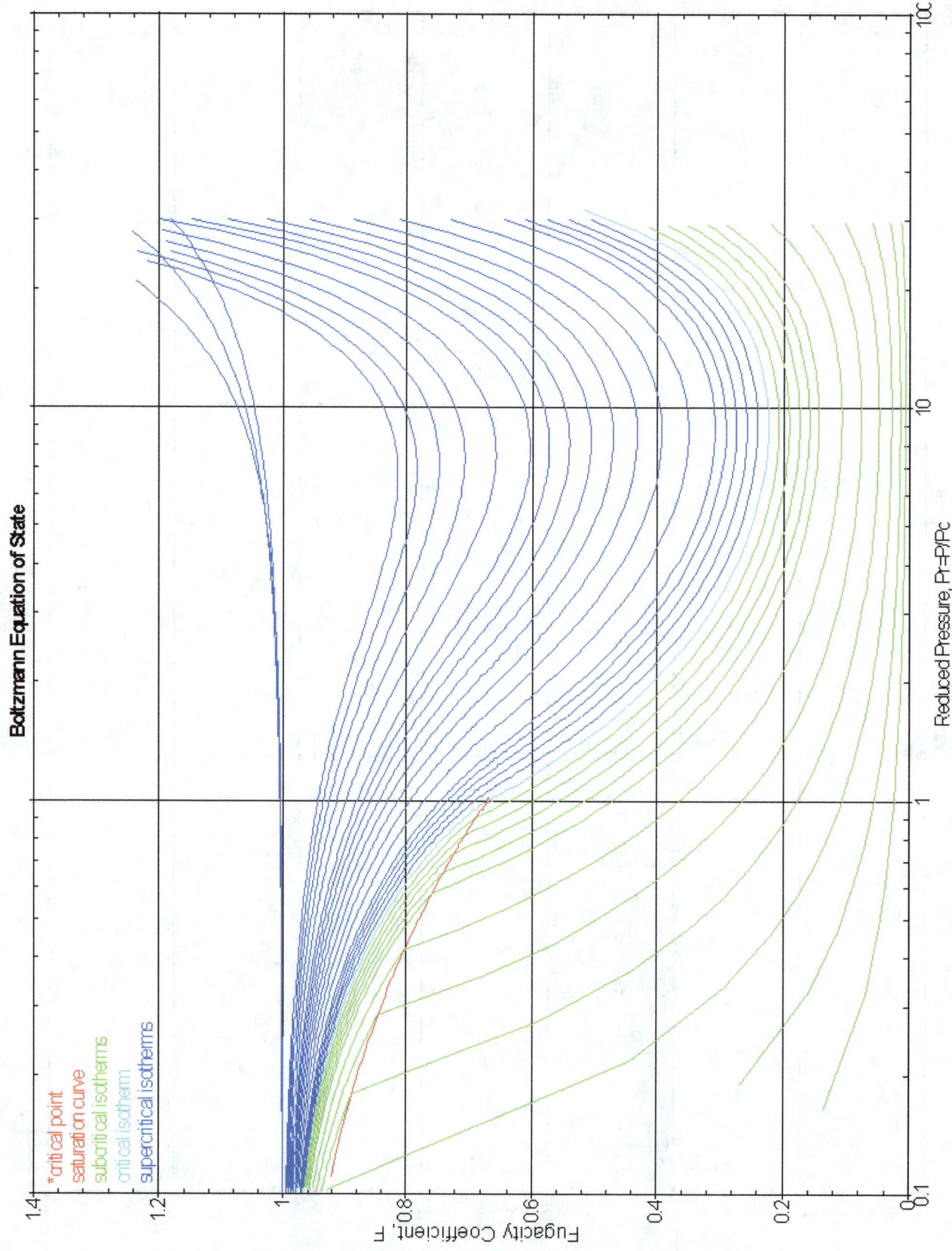

Figure 55. Fugacity Coefficient Based on Boltzmann

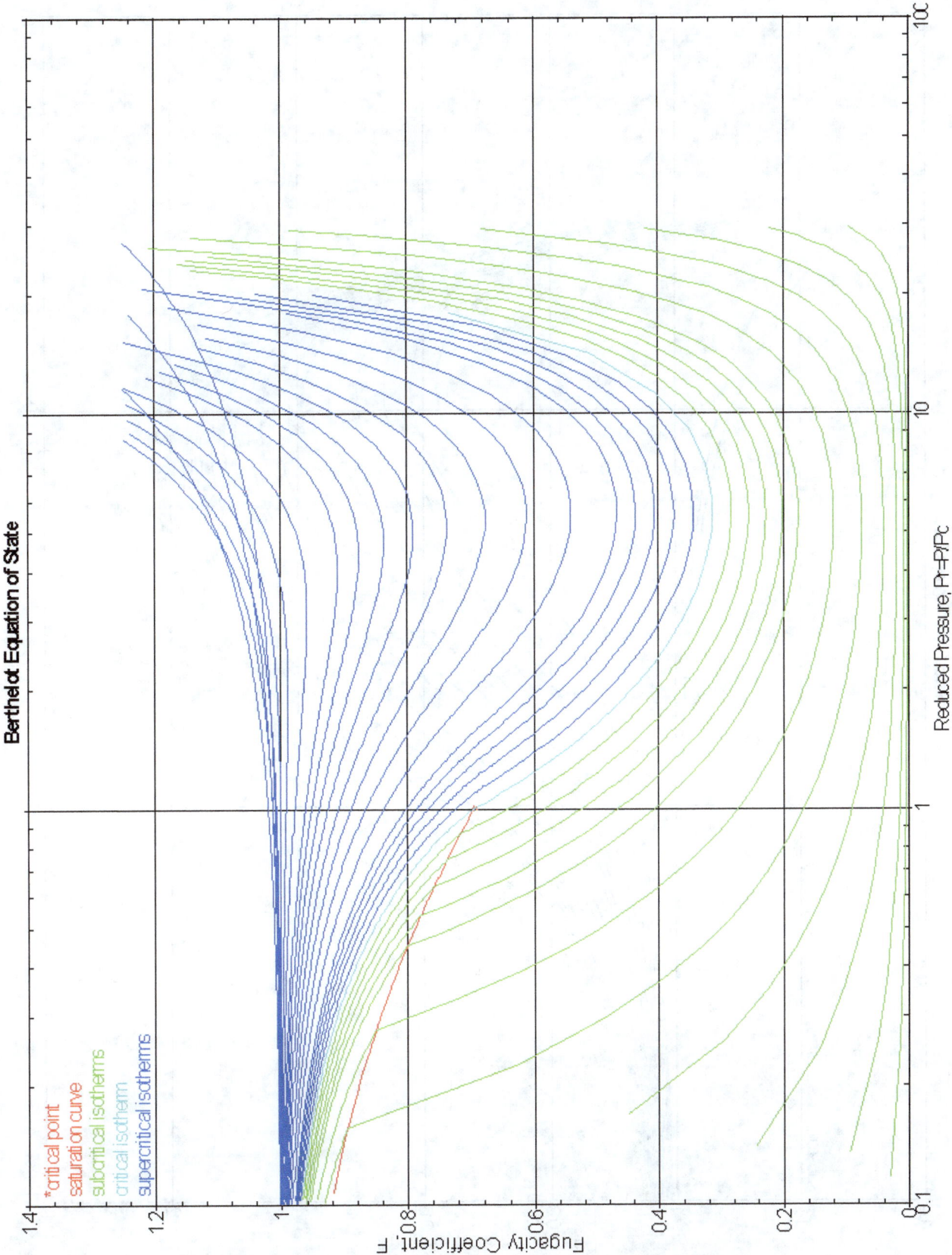

Figure 56. Fugacity Coefficient Based on Berthelot

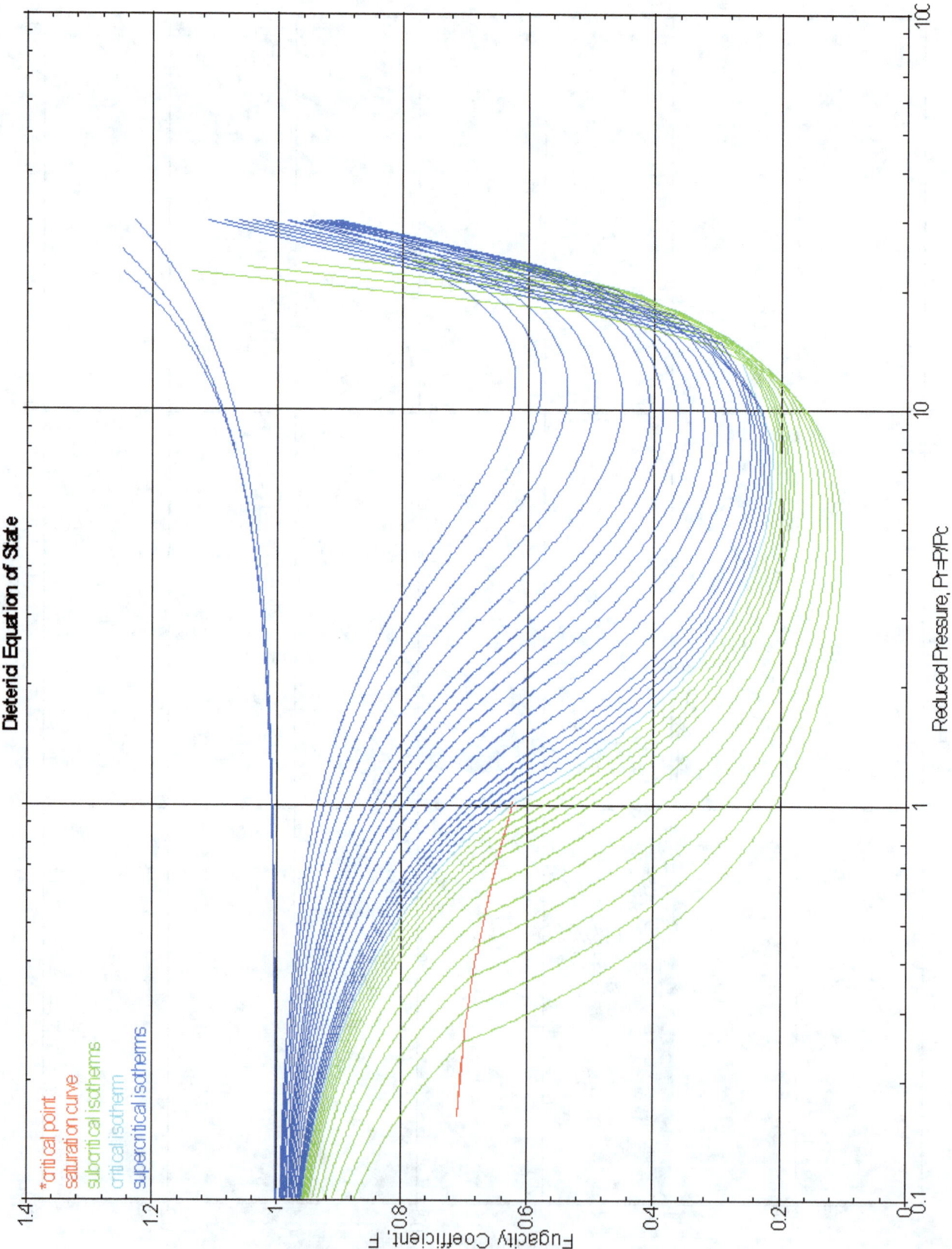

Figure 57. Fugacity Coefficient Based on Dieterici

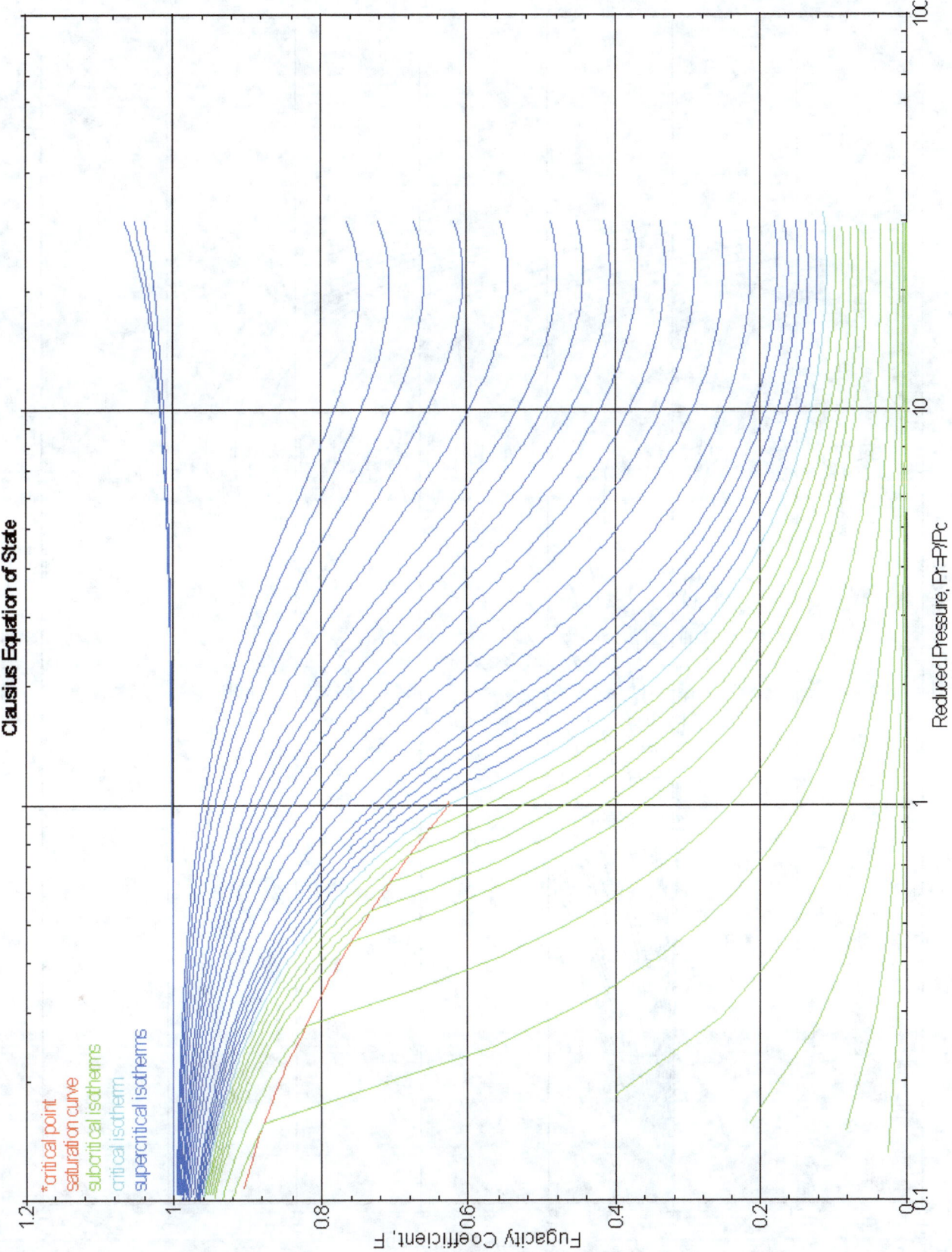

Figure 58. Fugacity Coefficient Based on Clausius

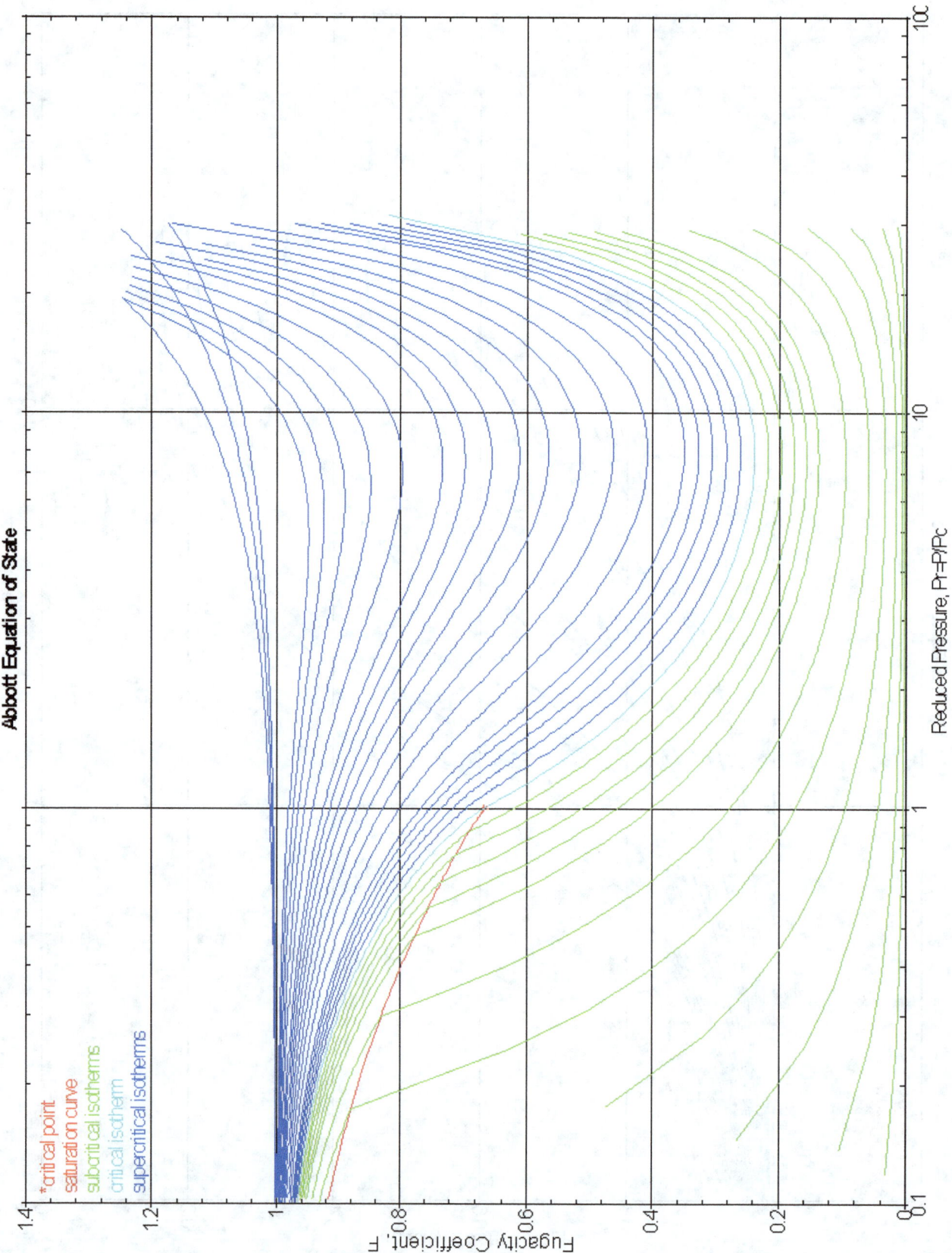

Figure 59. Fugacity Coefficient Based on Abbott's Modification

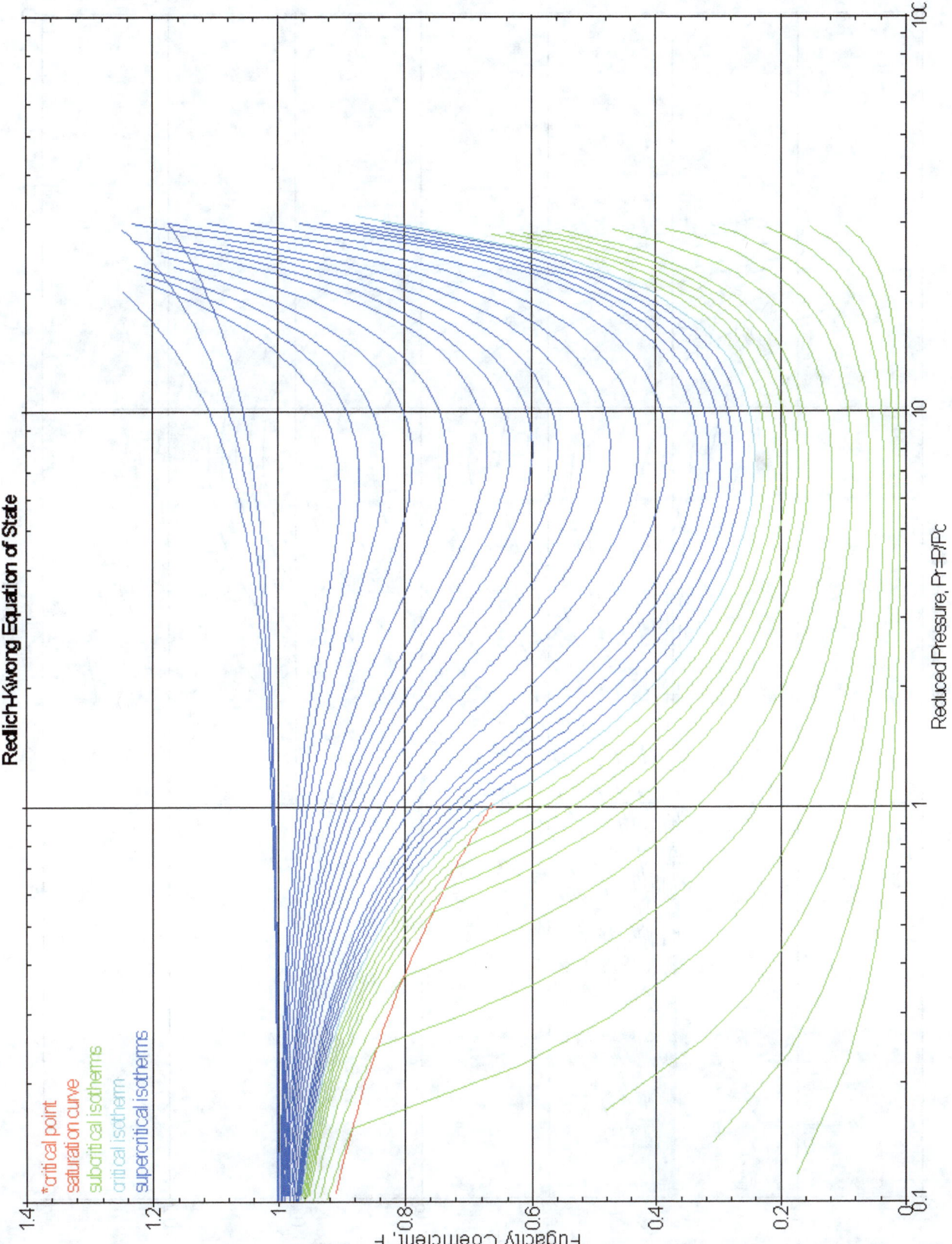

Figure 60. Fugacity Coefficient Based on Redlich-Kwong

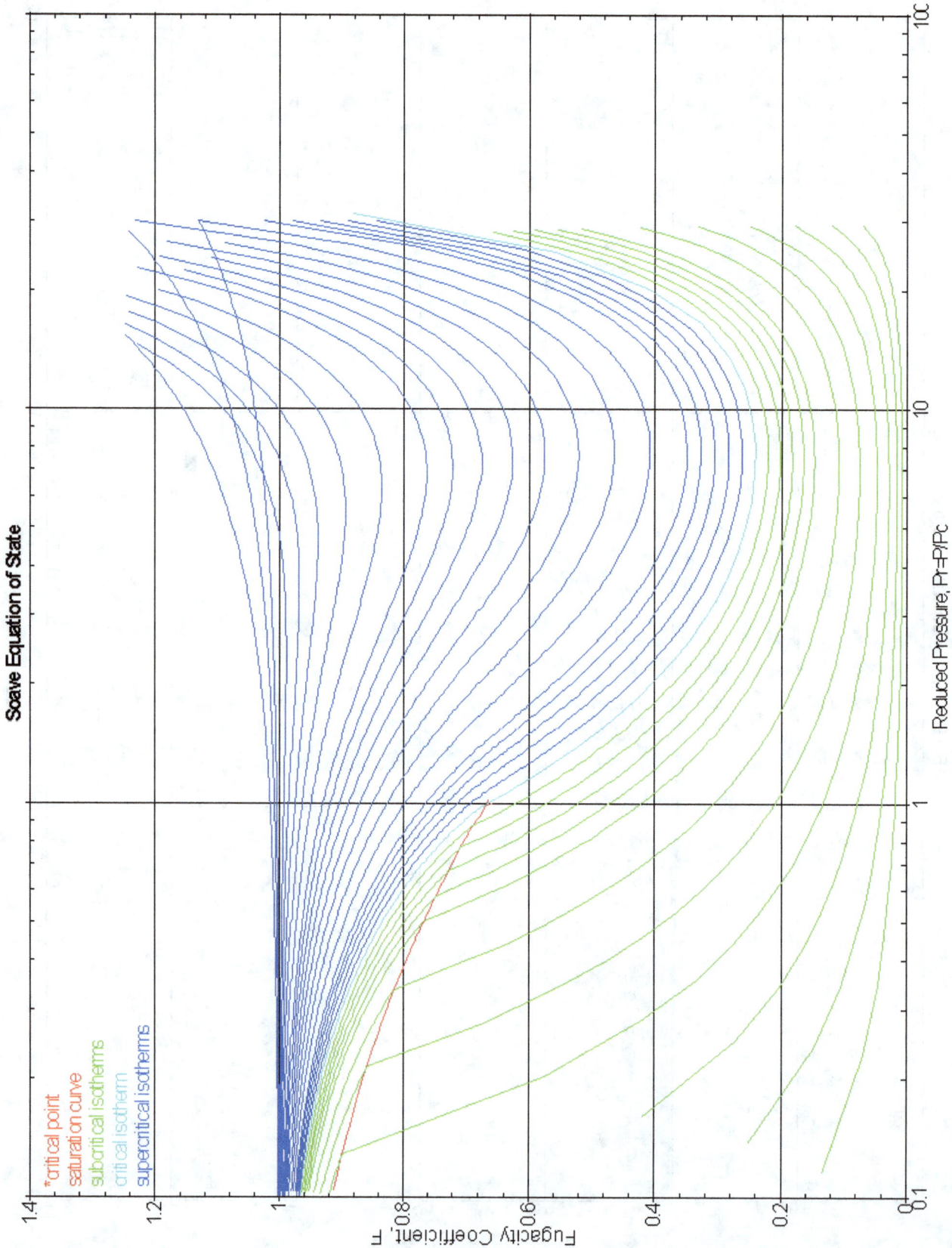

Figure 61. Fugacity Coefficient Based on Soave's Modification

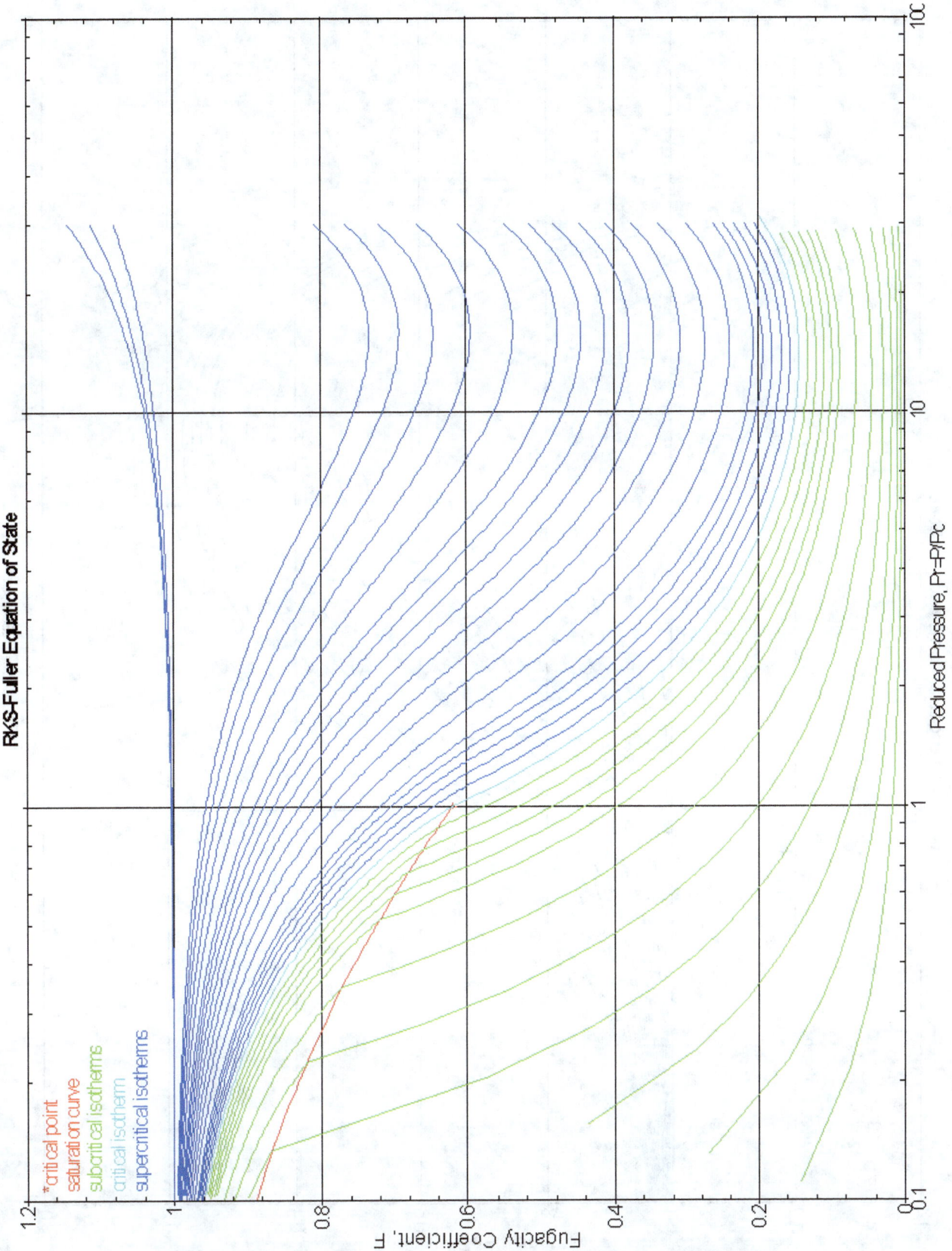

Figure 62. Fugacity Coefficient Based on Fuller's Modification

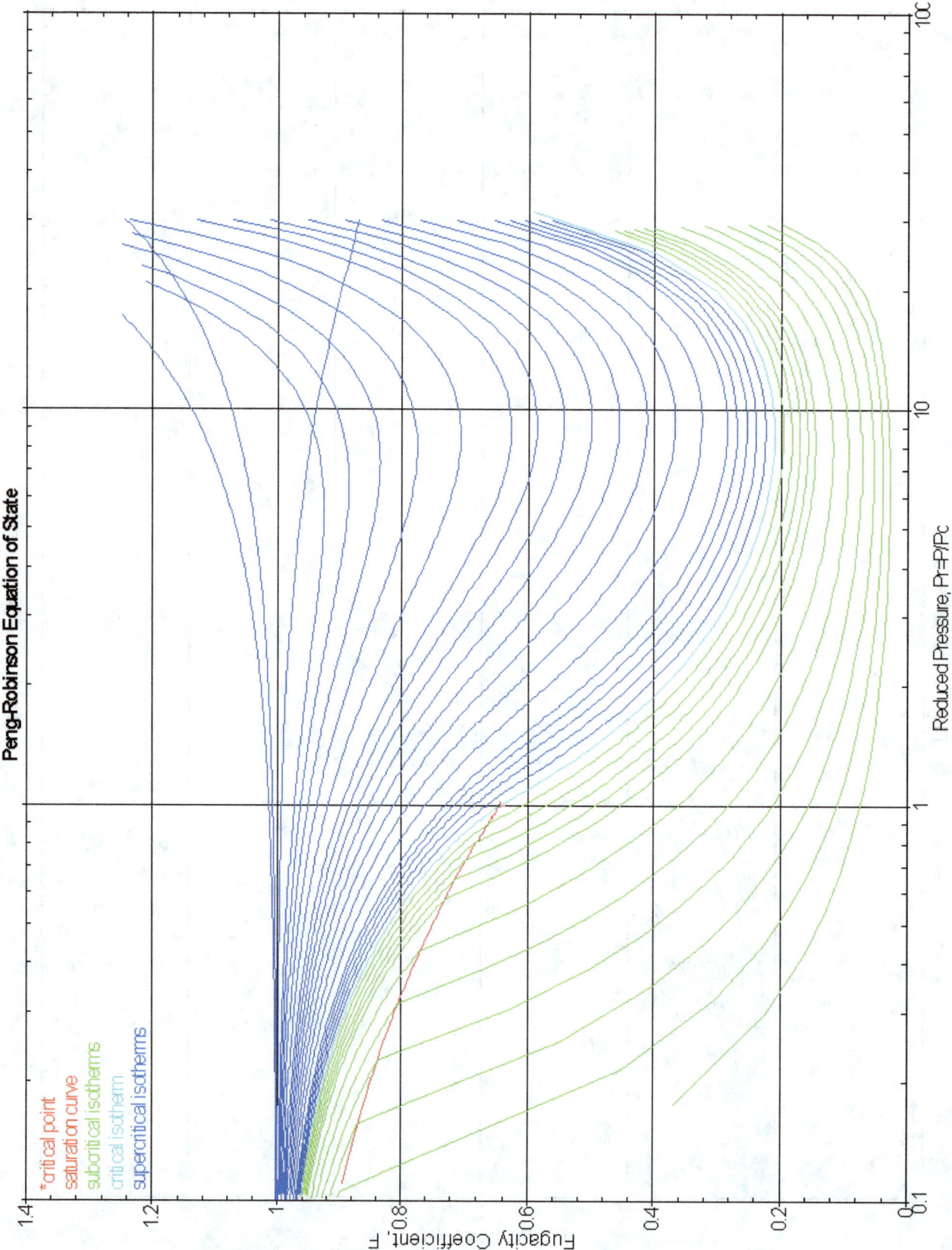

Figure 63. Fugacity Coefficient Based on Peng-Robinson

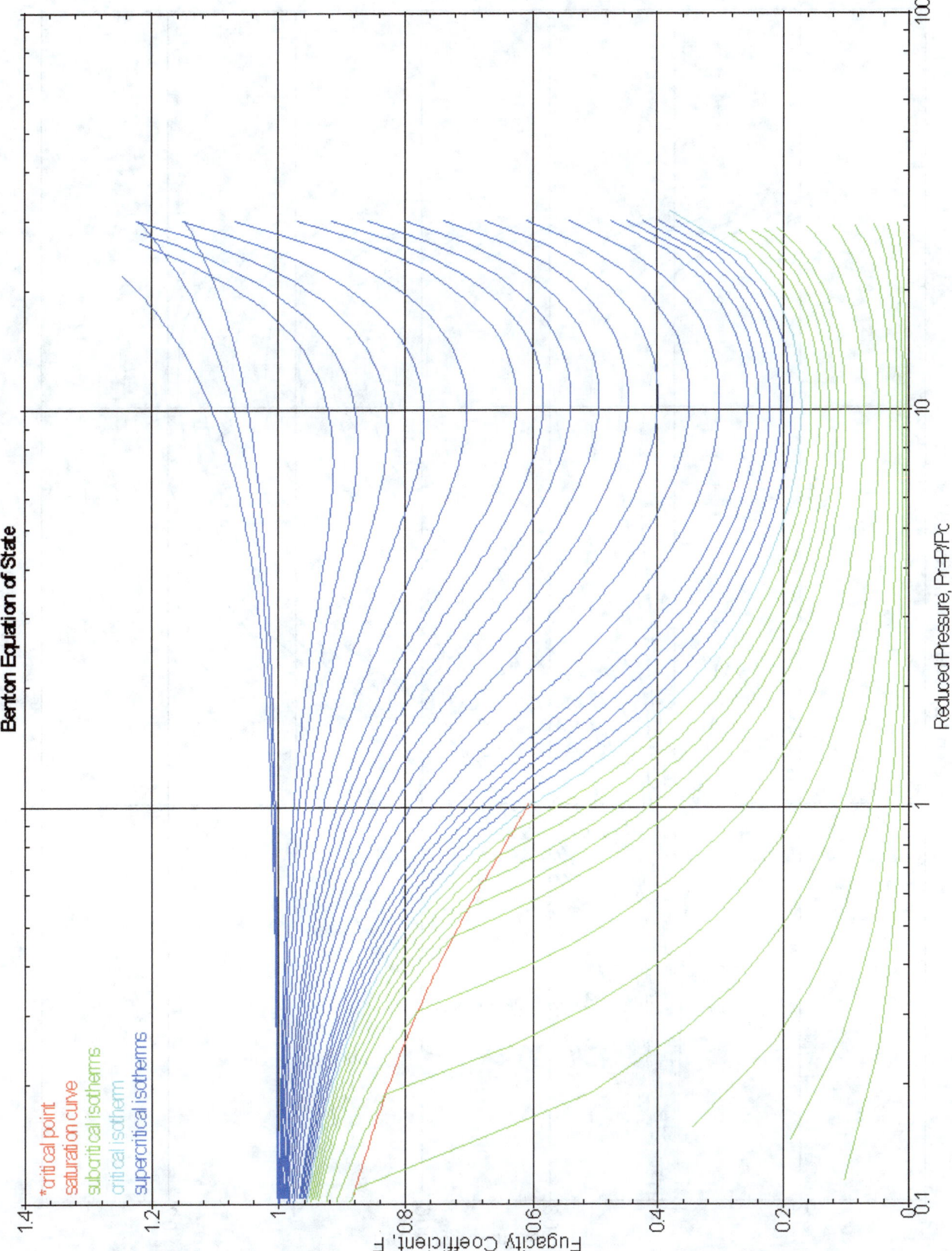

Figure 64. Fugacity Coefficient Based on Author's Modification

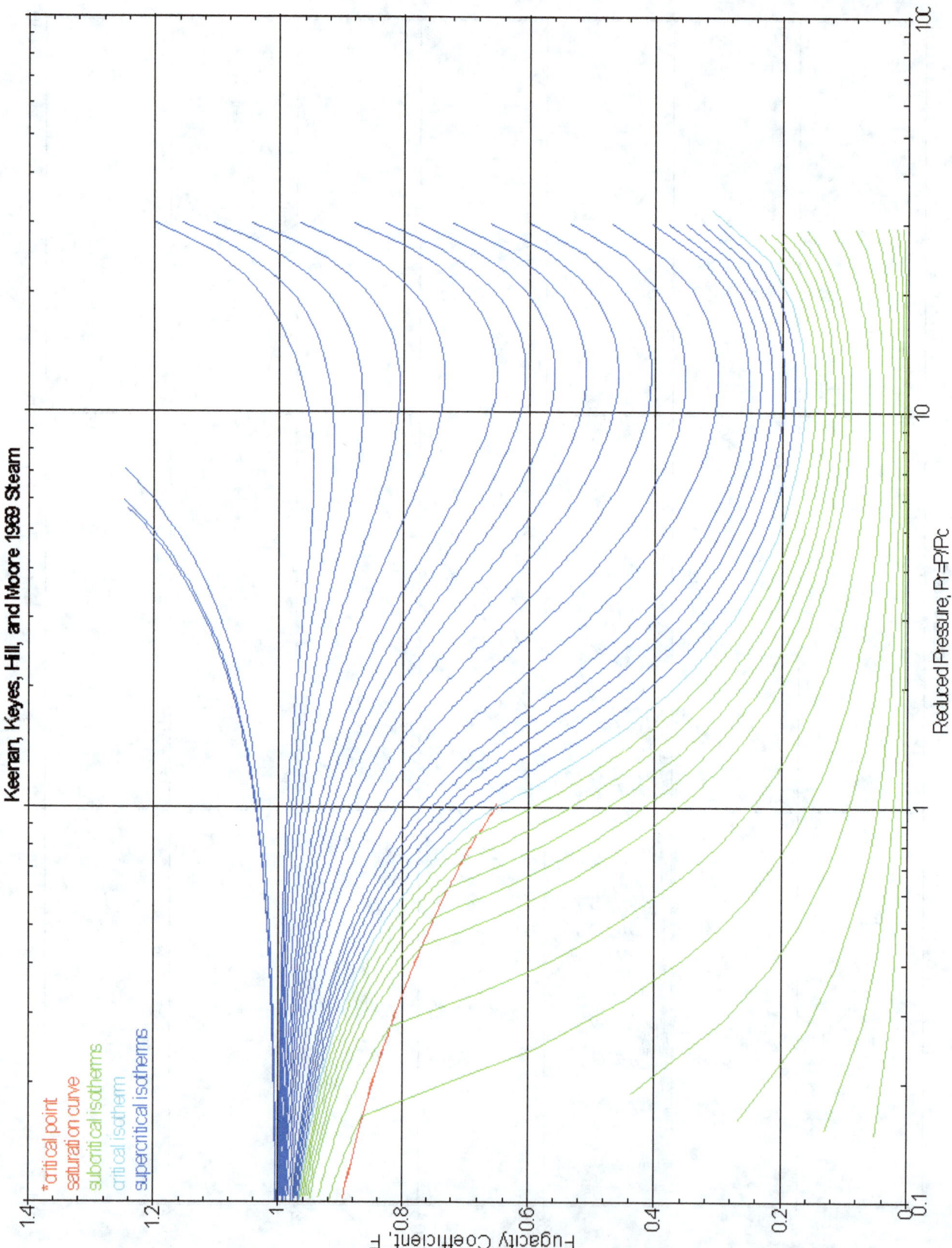

Figure 65. Fugacity Coefficient Based on Keenan, Keyes, Hill, and Moore

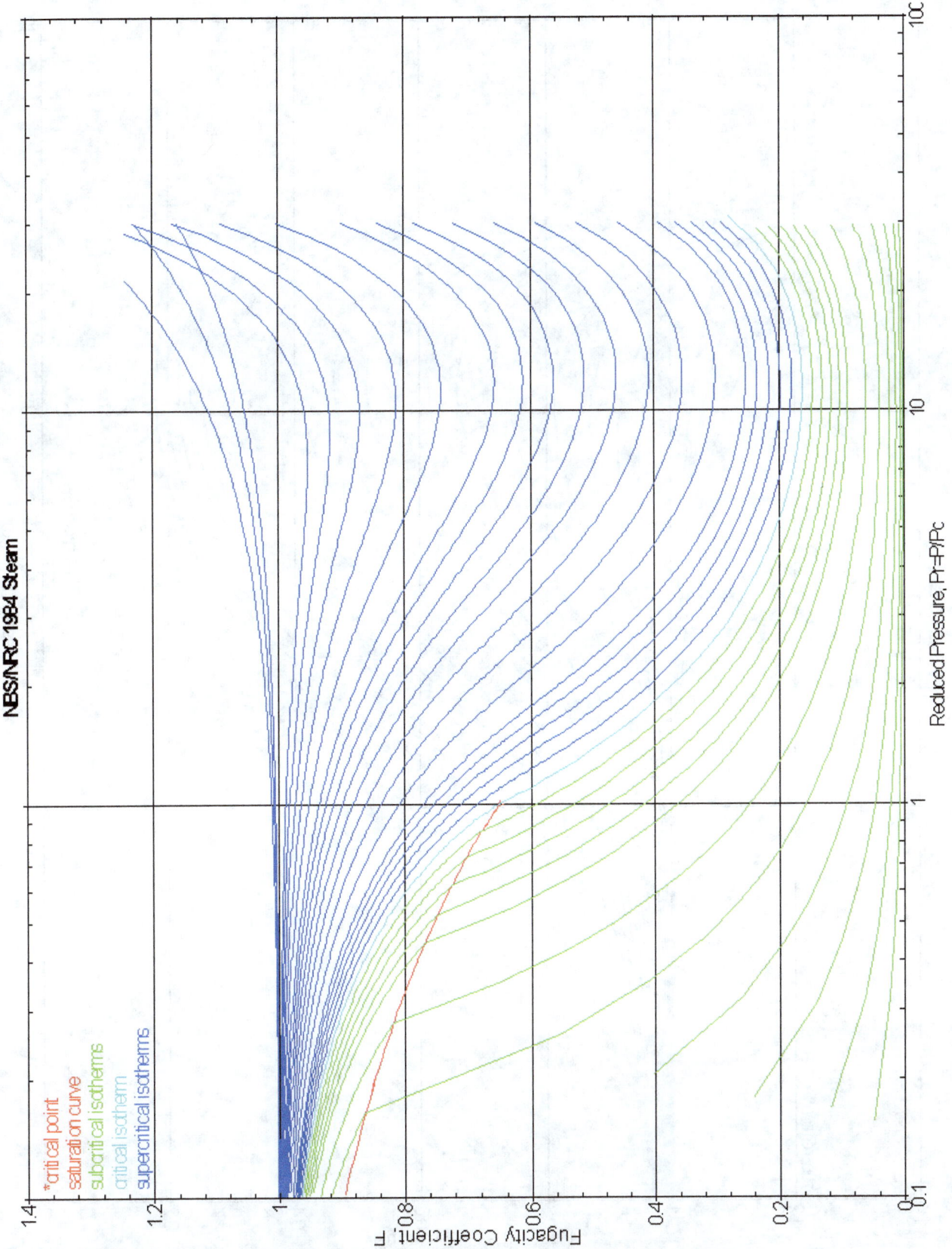

Figure 66. Fugacity Coefficient Based on Haar, Gallagher, and Kell

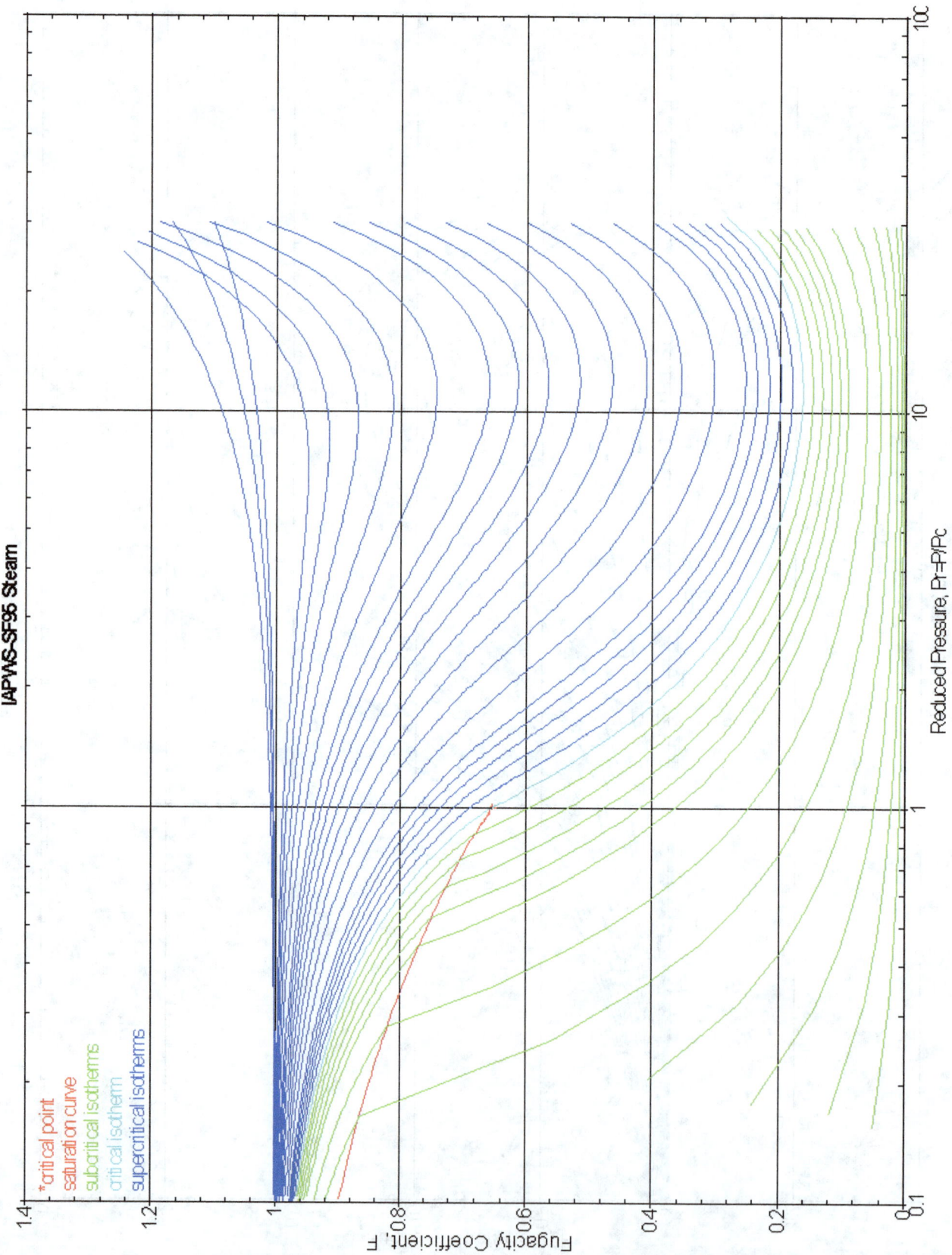

Figure 67. Fugacity Coefficient Based on Wagner and Pruß

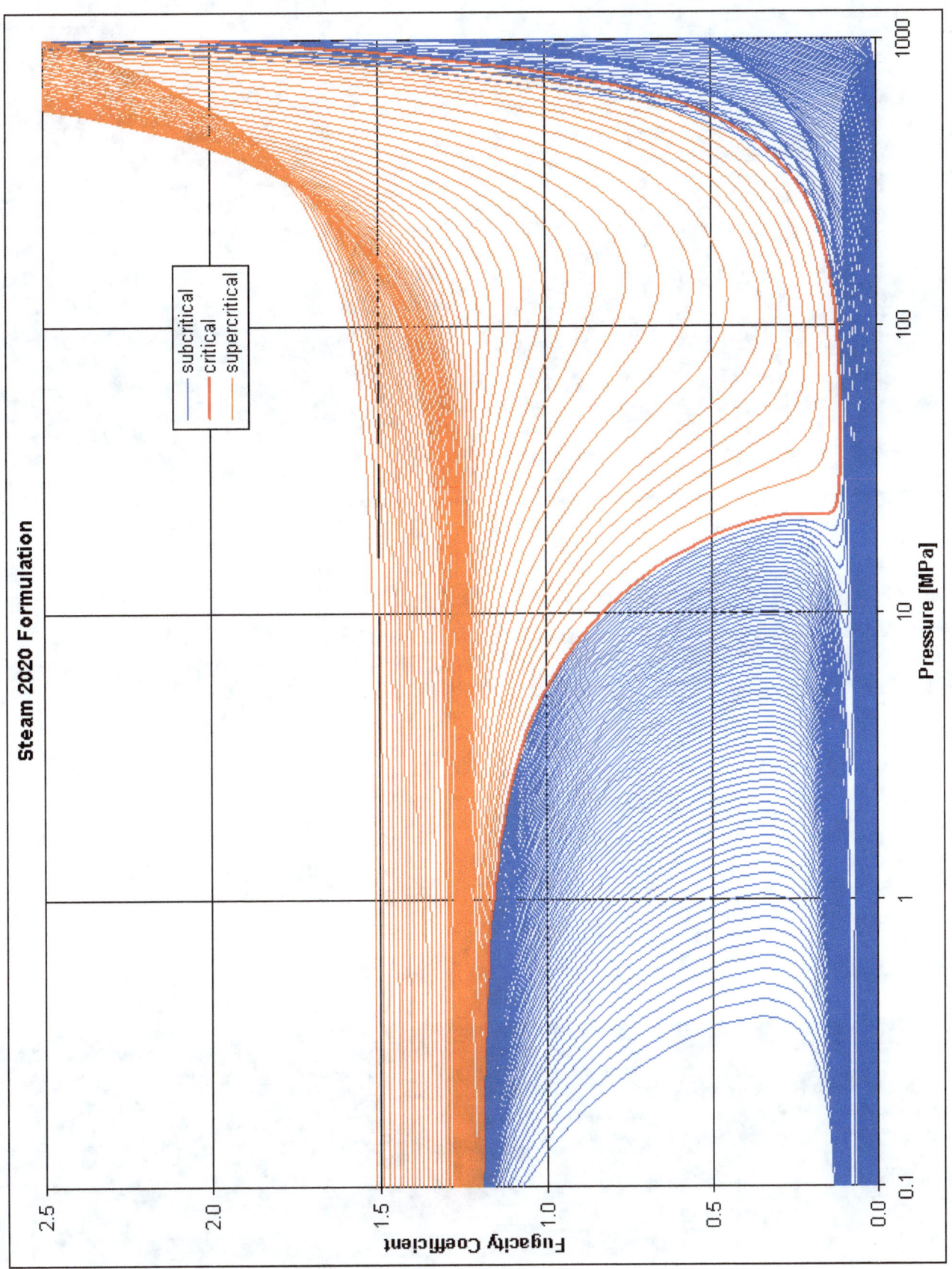

Figure 68. Fugacity Coefficient Based on Steam 2020 Formulation

Note that this figure also shows the metastable regions, which the others do not.

Chapter 6. Residual Enthalpy

Residual enthalpy is the departure from ideal gas behavior. All of the steam formulations discussed in this text begin with the ideal gas behavior plus terms to handle the departure or the difference between real and ideal. Residual enthalpy is one of these, as defined by the following:

$$\frac{H_0 - H}{RT_C} = T_R(1-Z) - Z_C \int_\infty^{V_R} \left(P_R - T_R \frac{\partial P_R}{\partial T_R} \right) dV_R \qquad (6.1)$$

where H_0 is the ideal gas enthalpy and R is the ideal gas constant. There are other forms but this one has the integral in terms of specific volume (like we did for fugacity, integrating by parts), as the pressure formulation is mathematically useless. The residual enthalpy for the van der Waals and Boltzmann EOS is:

$$H_R = T_R(1-Z) + \frac{A}{V_R} \qquad (6.2)$$

The residual enthalpy for the Berthelot EOS is:

$$H_R = T_R(1-Z) + \frac{2A}{T_R V_R} \qquad (6.3)$$

The residual enthalpy for the Clausius EOS is:

$$H_R = T_R(1-Z) + \frac{2A}{T_R(V_R + C)} \qquad (6.4)$$

The other formulations require more extensive formulas and in some cases (e.g., Dieterici) numerical integration. As before, we begin with the Nelson-Obert data.

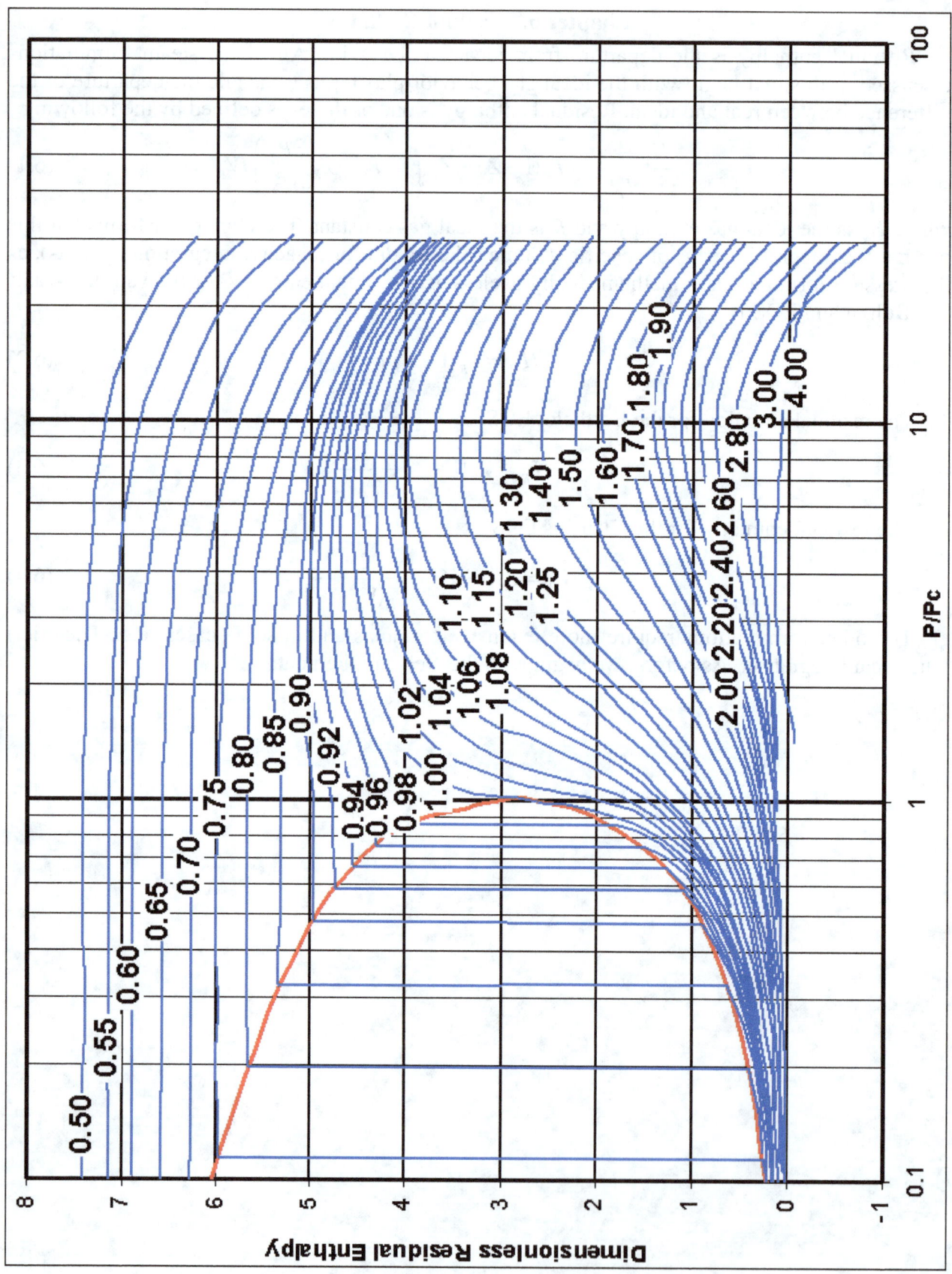

Figure 69. Residual Enthalpy Based on Nelson-Obert

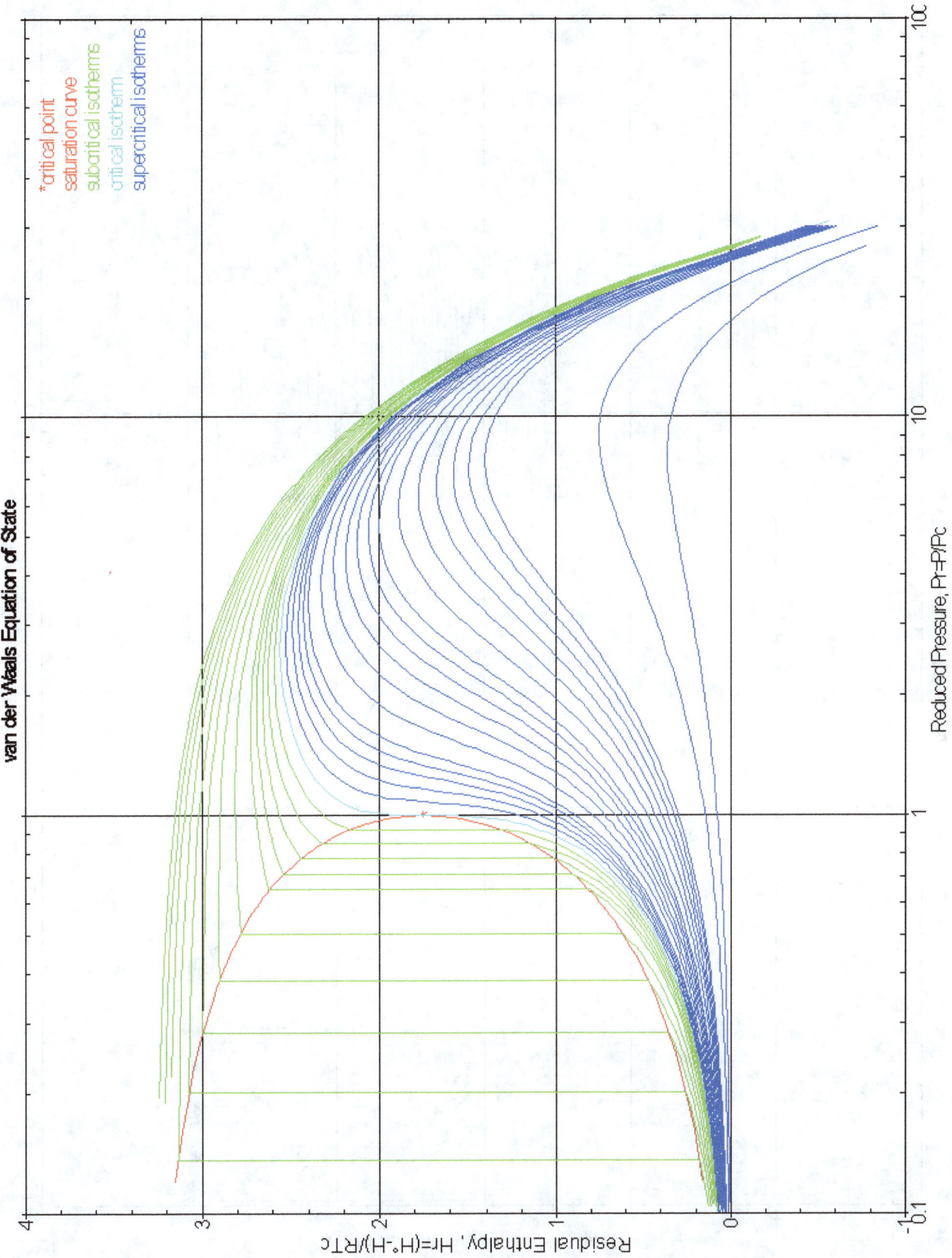

Figure 70. Residual Enthalpy Based on van der Waals

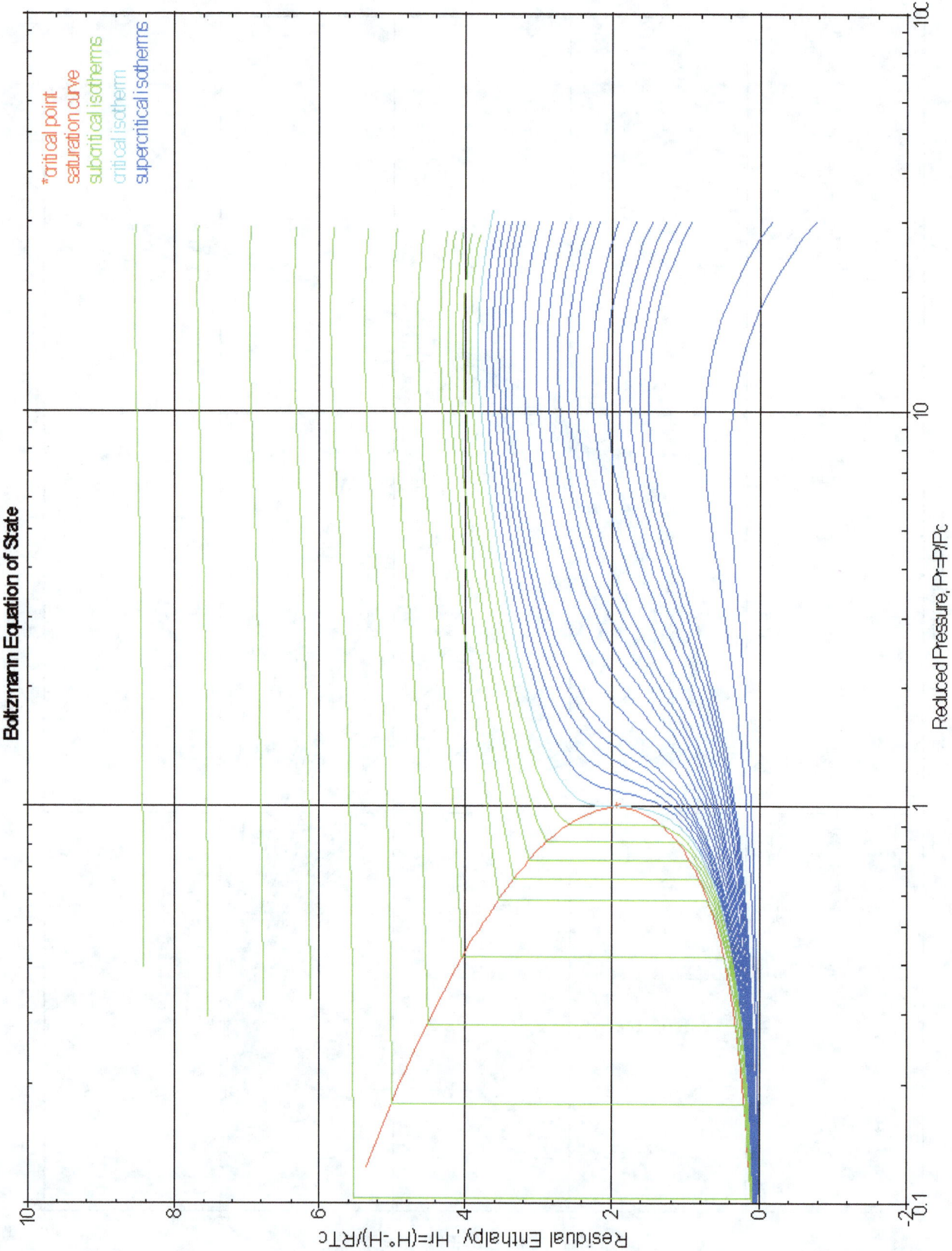

Figure 71. Residual Enthalpy Based on Boltzmann

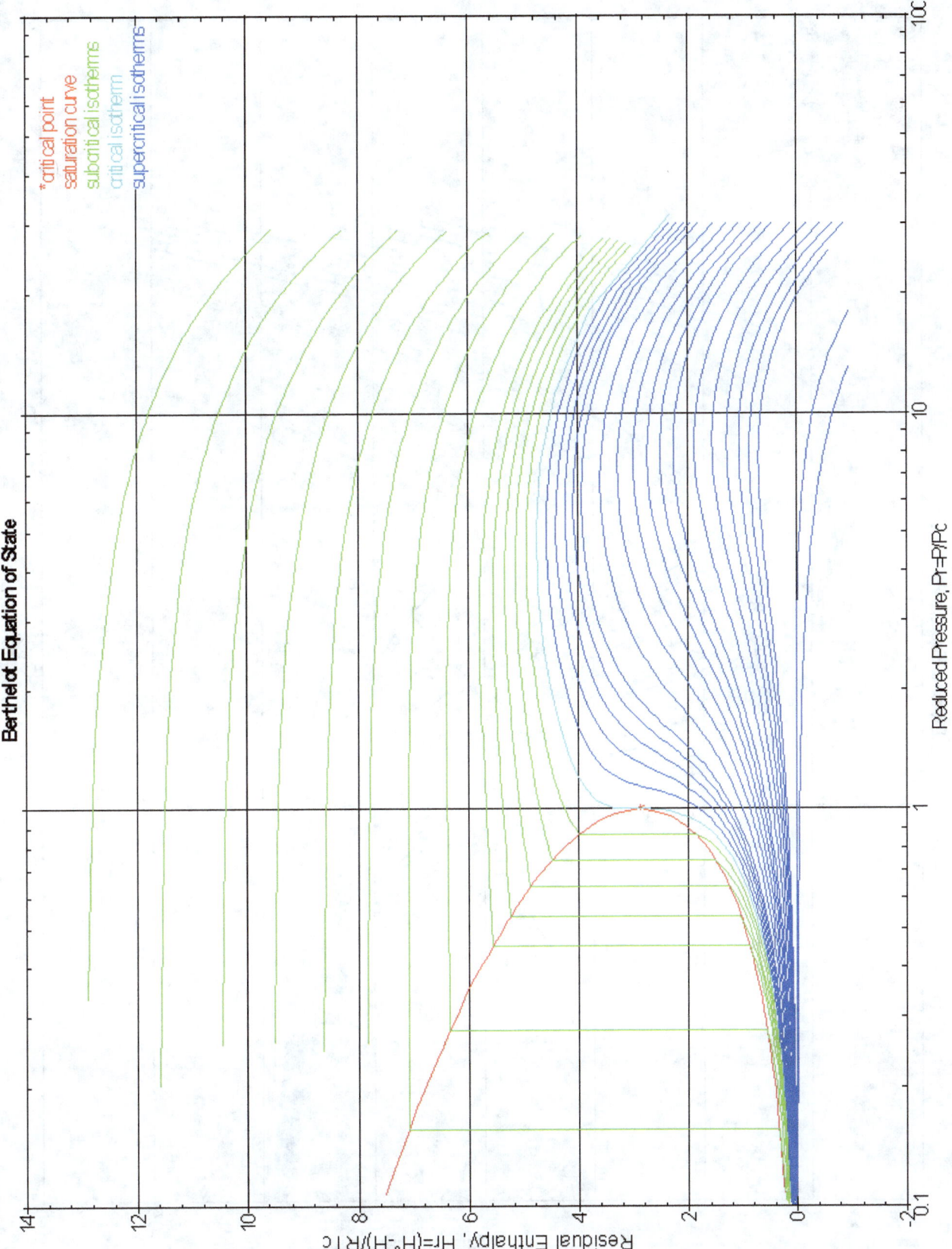

Figure 72. Residual Enthalpy Based on Berthelot

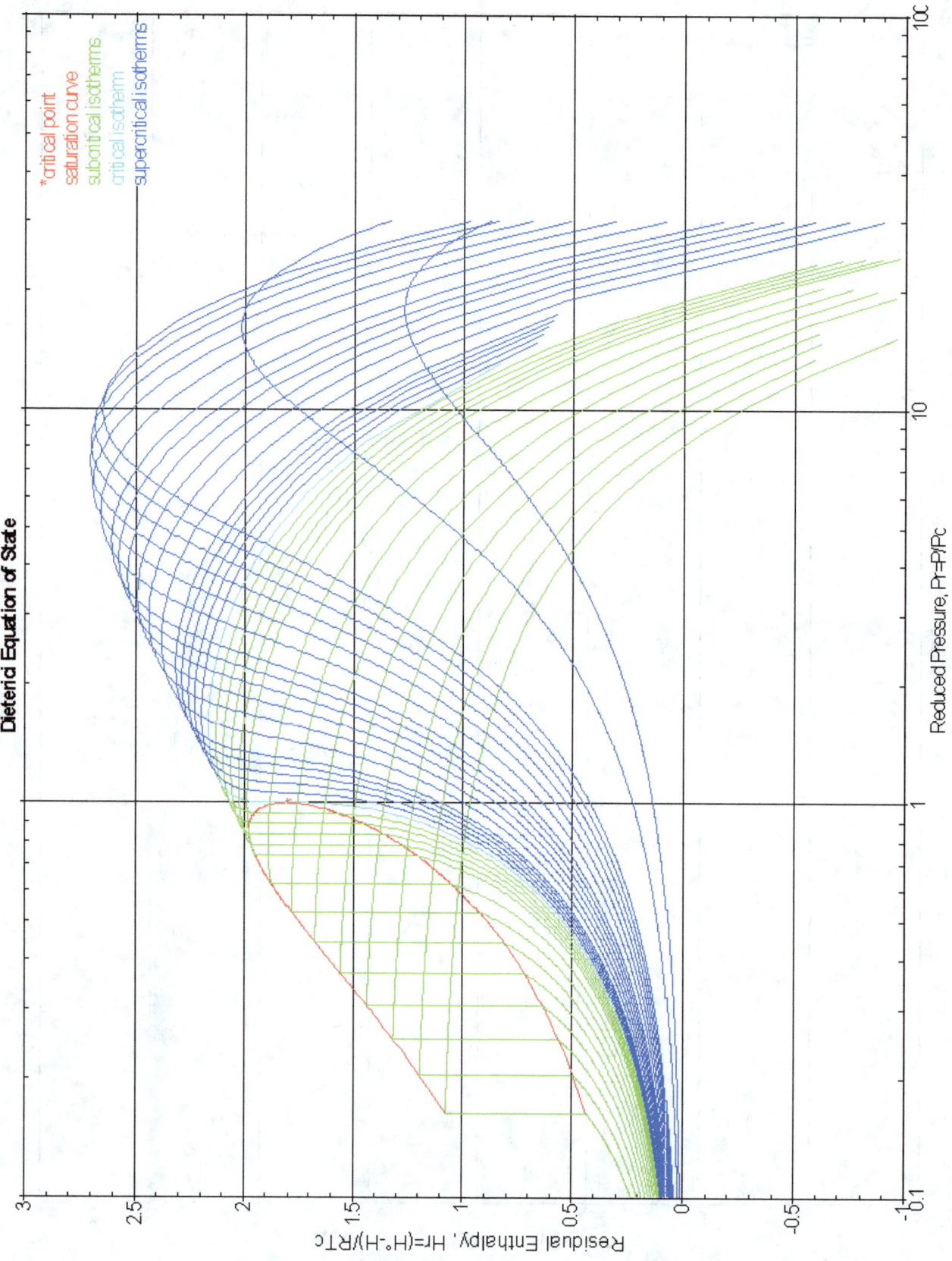

Figure 73. Residual Enthalpy Based on Dieterici

The behavior illustrated in the figure above is unacceptable.

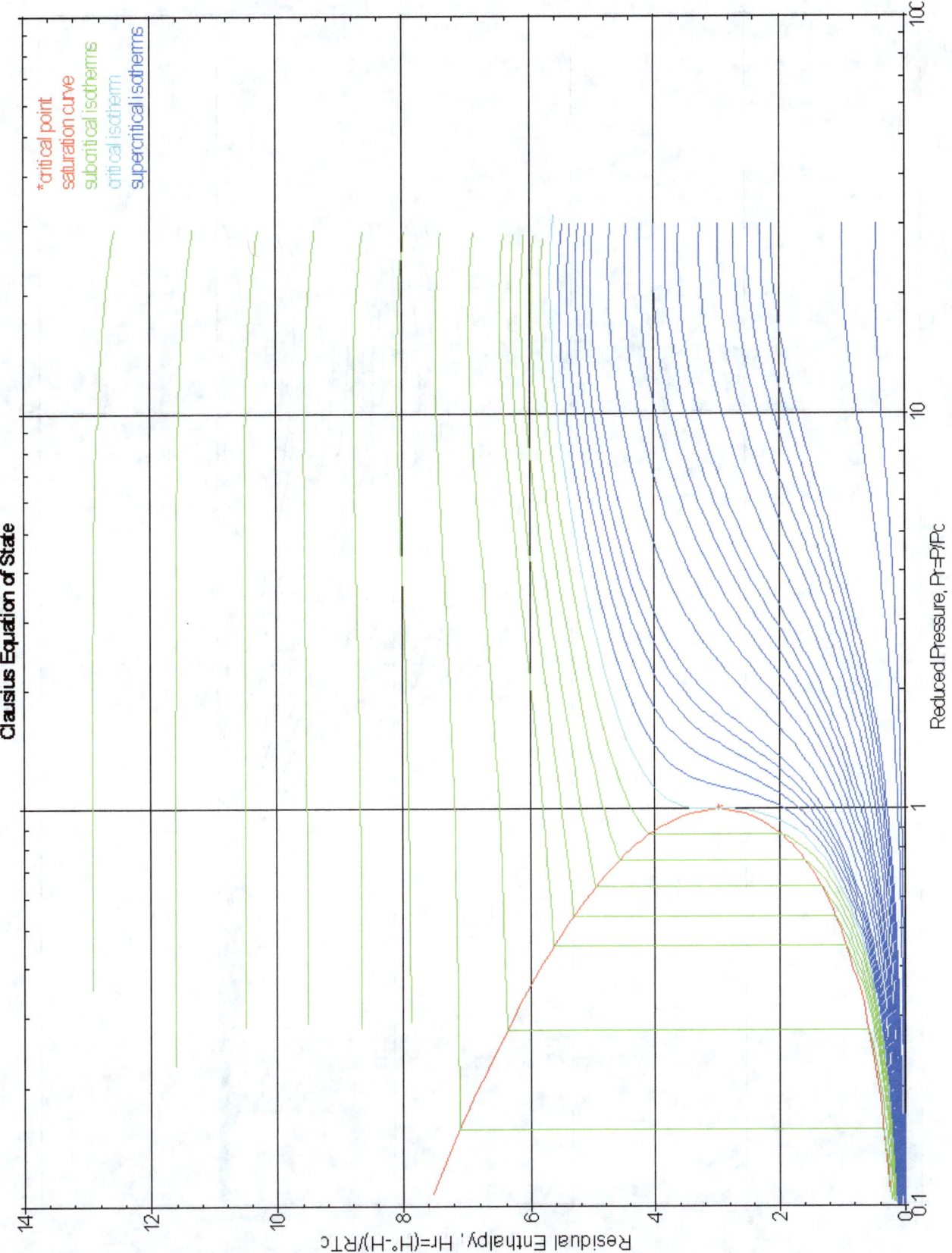

Figure 74. Residual Enthalpy Based on Clausius

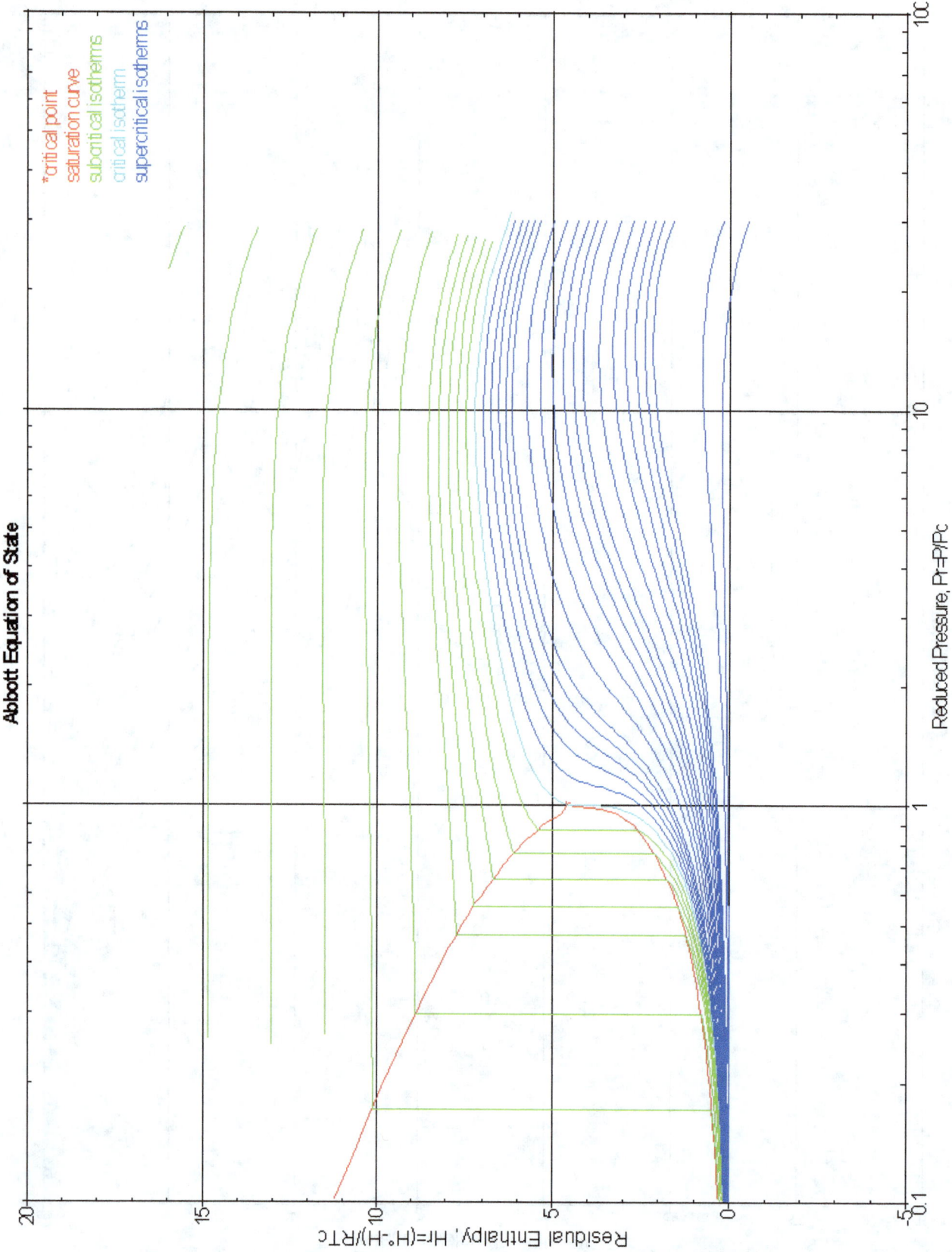

Figure 75. Residual Enthalpy Based on Abbott's Modification

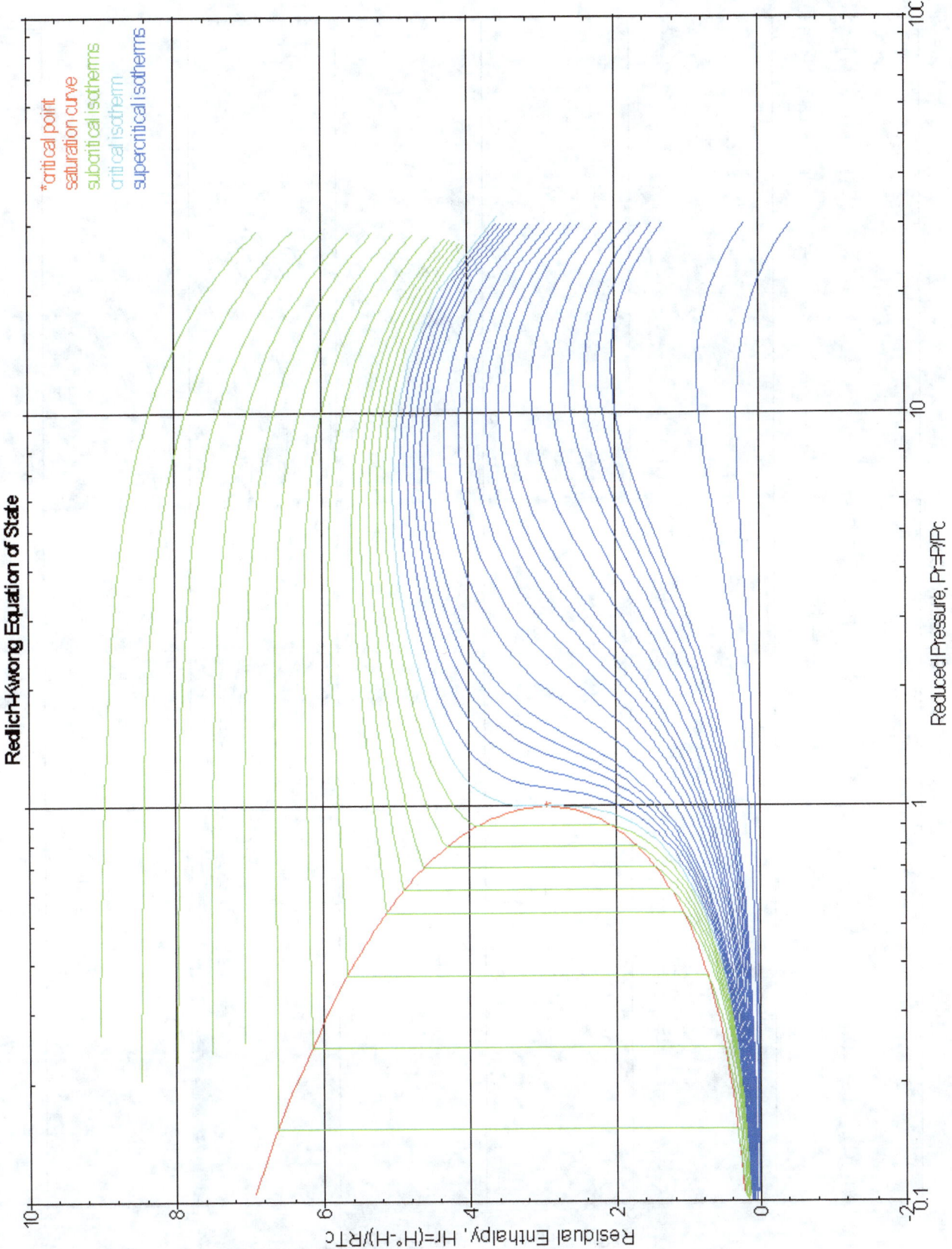

Figure 76. Residual Enthalpy Based on Redlich-Kwong

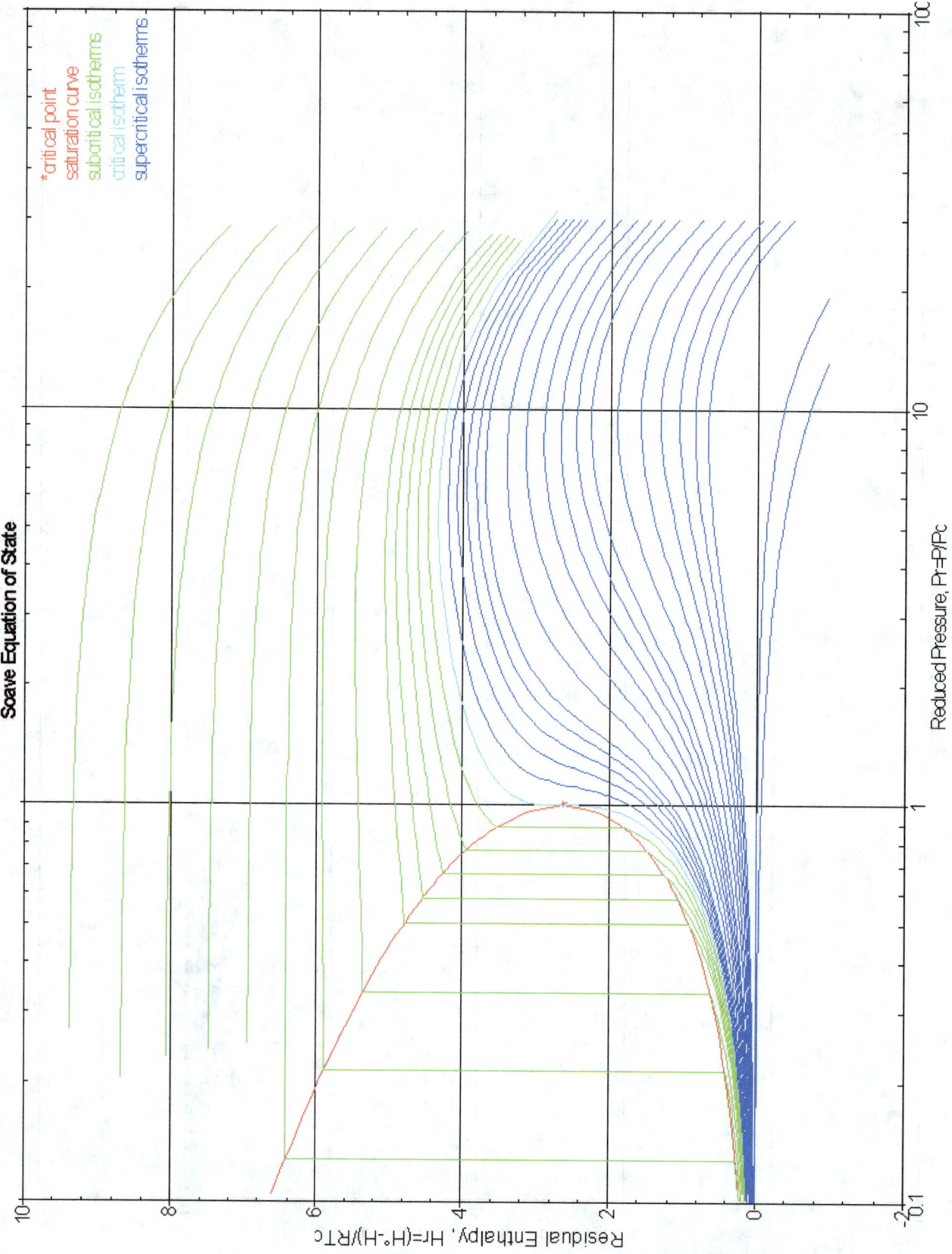

Figure 77. Residual Enthalpy Based on Soave's Modification

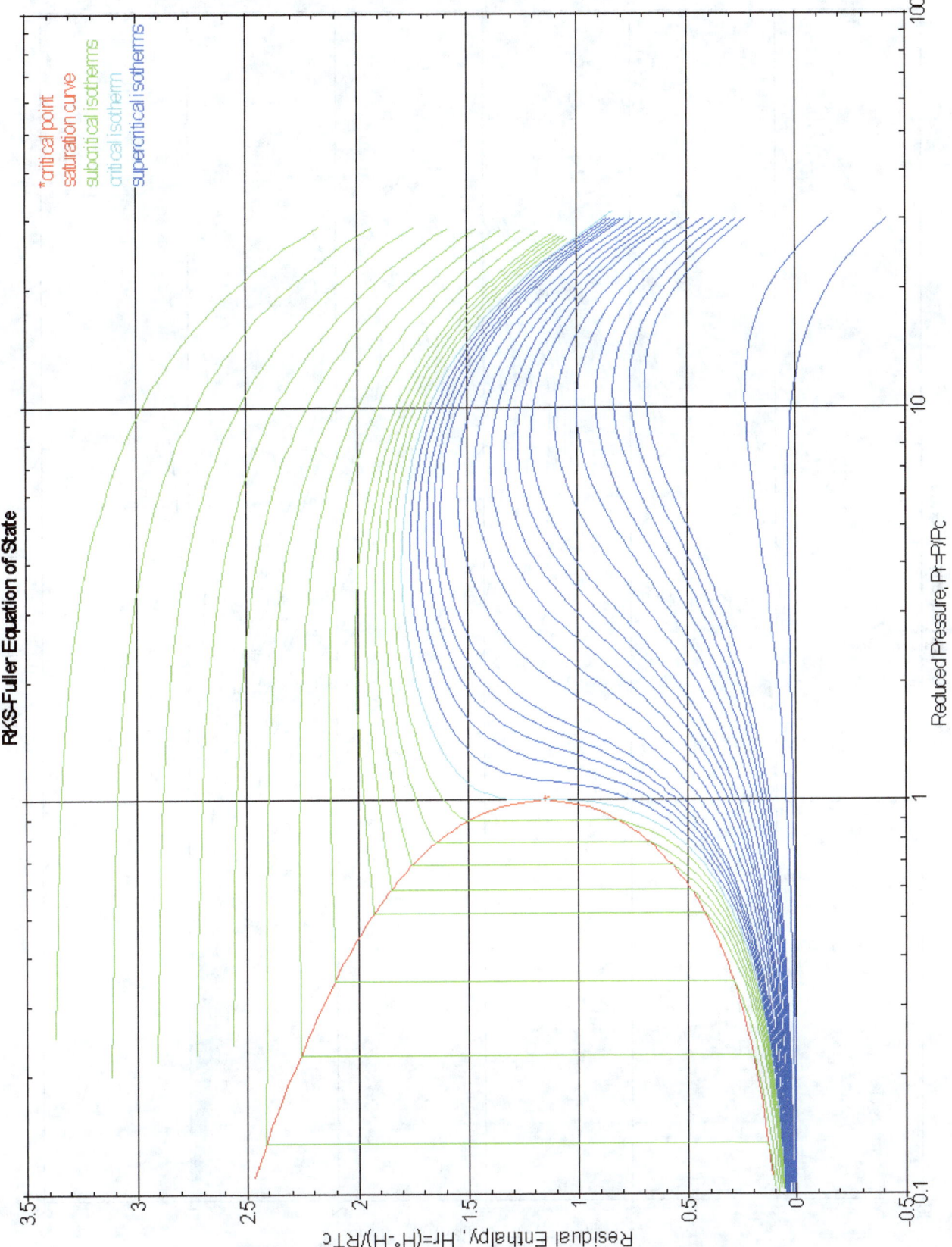

Figure 78. Residual Enthalpy Based on Fuller's Modification

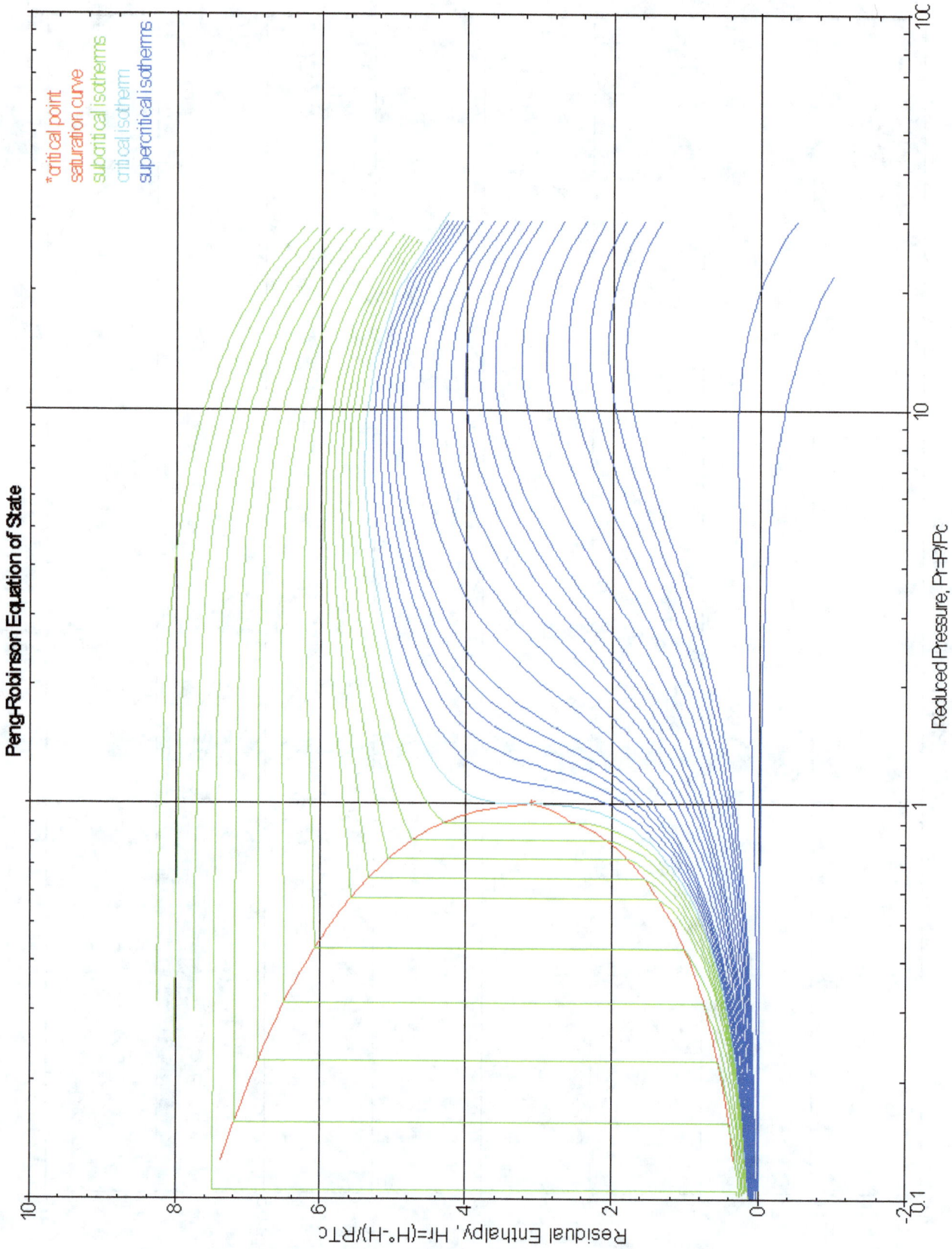

Figure 79. Residual Enthalpy Based on Peng-Robinson

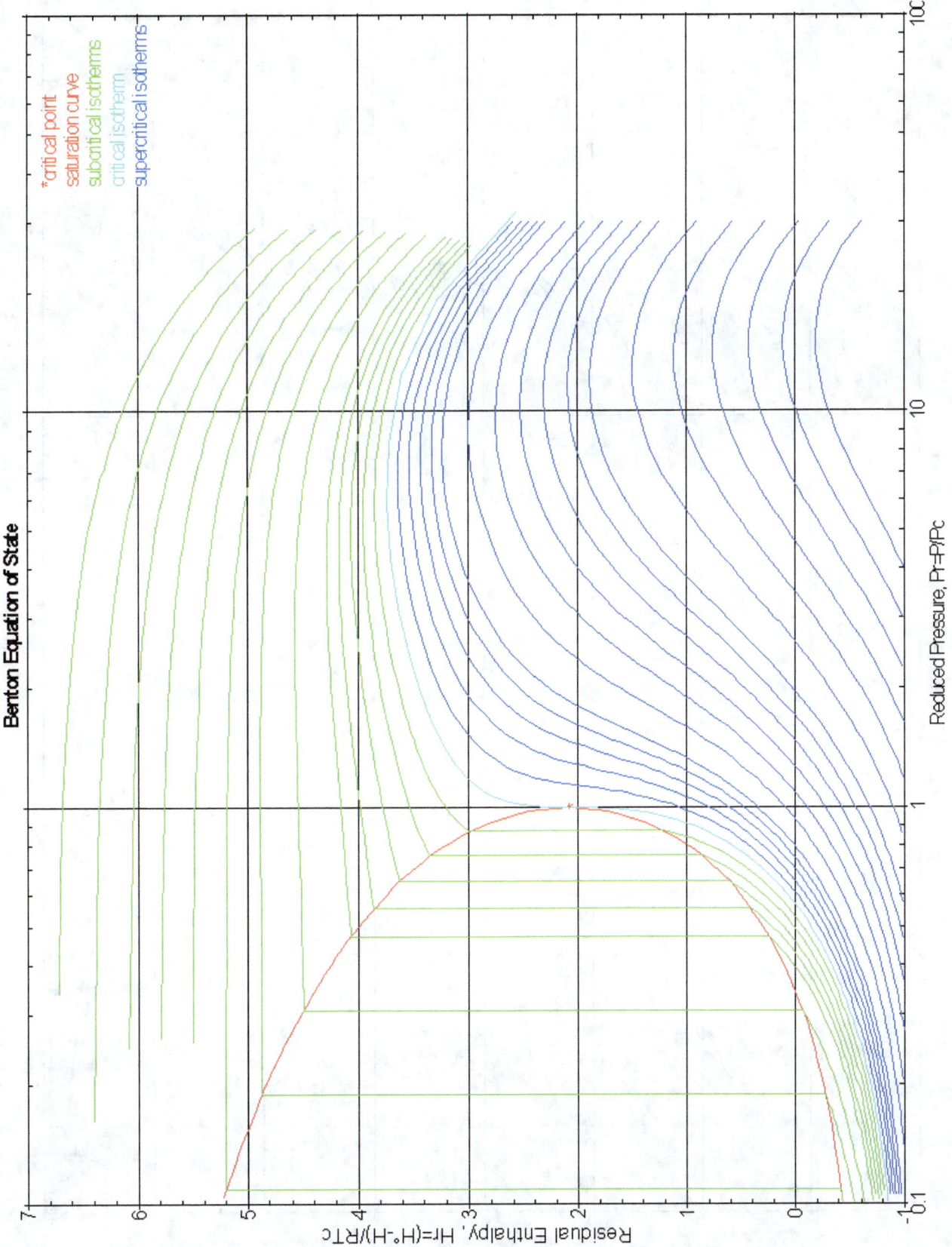

Figure 80. Residual Enthalpy Based on Author's Modification

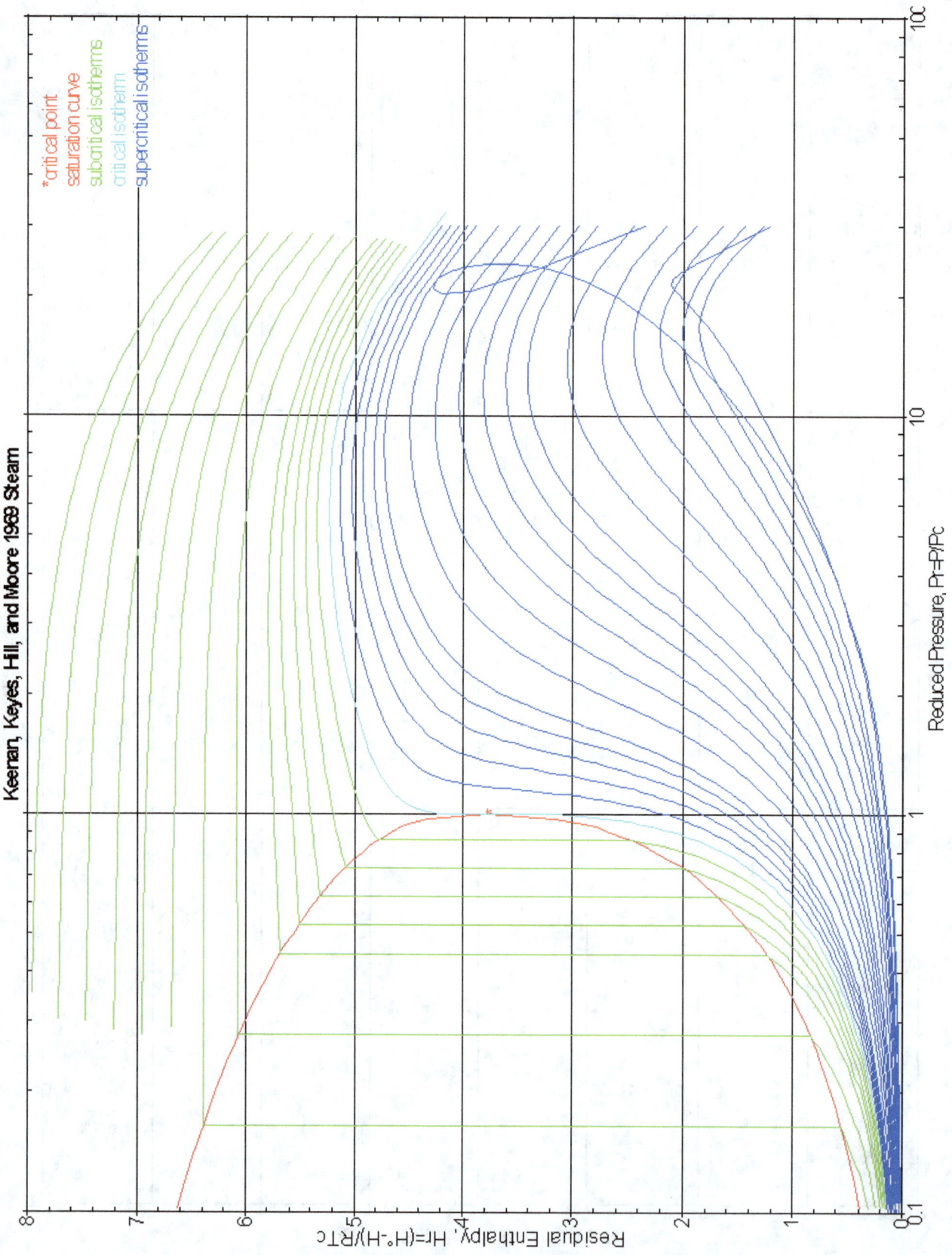

Figure 81. Residual Enthalpy Based on Keenan, Keyes, Hill, and Moore

The behavior at very high pressures shown above is wrong but beyond the recommended pressure limit for this formulation.

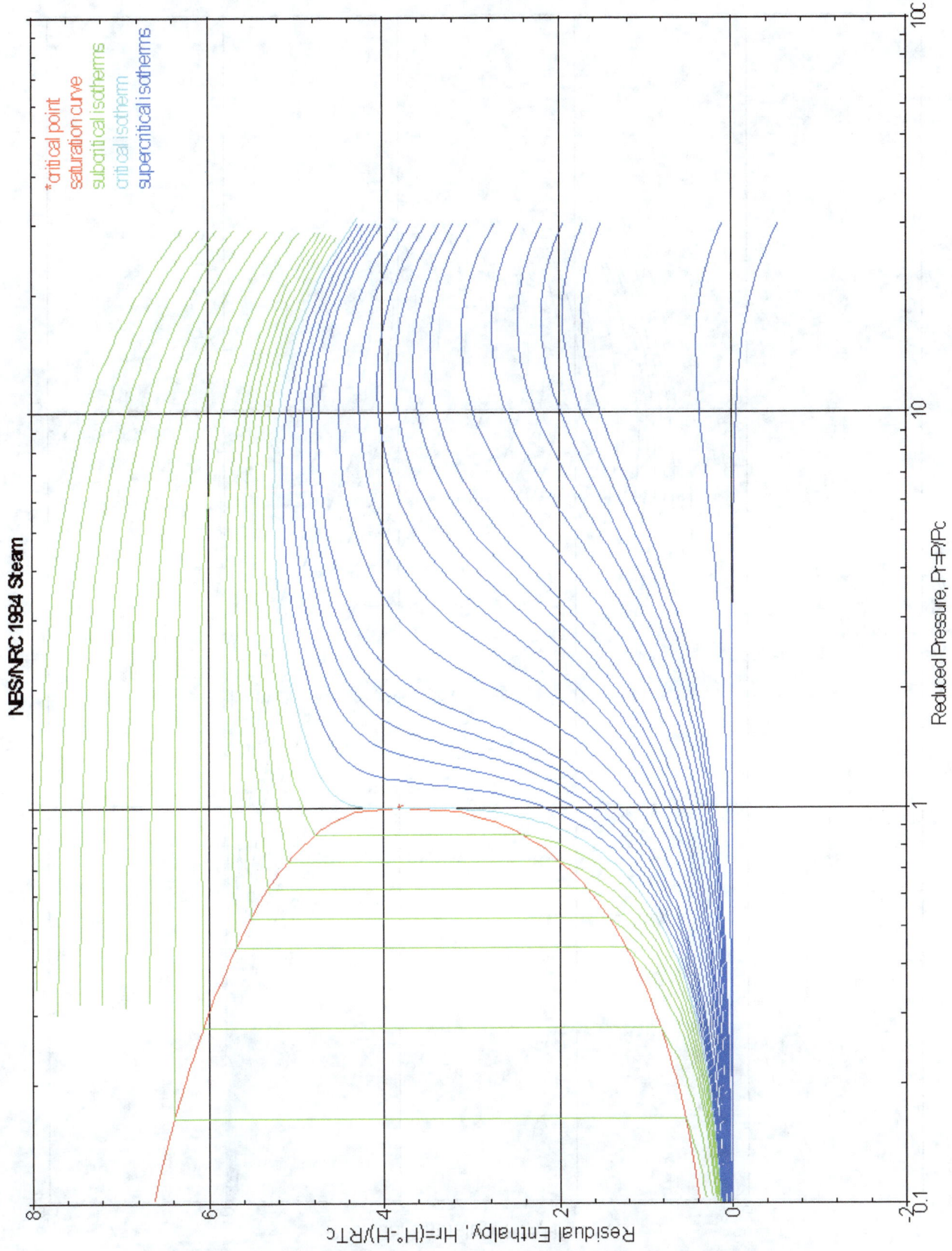

Figure 82. Residual Enthalpy Based on Haar, Gallagher, and Kell

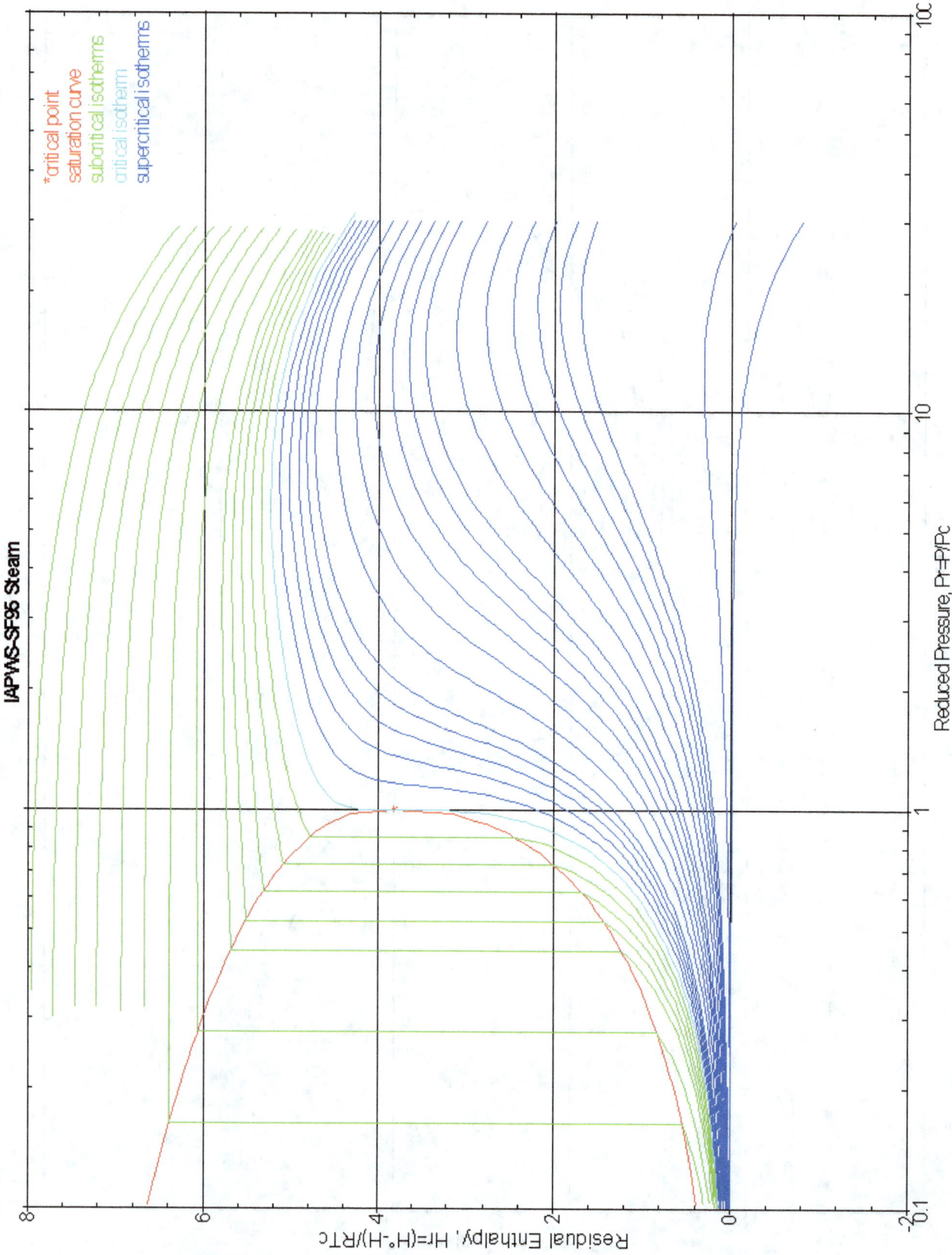

Figure 83. Residual Enthalpy Based on Wagner and Pruß

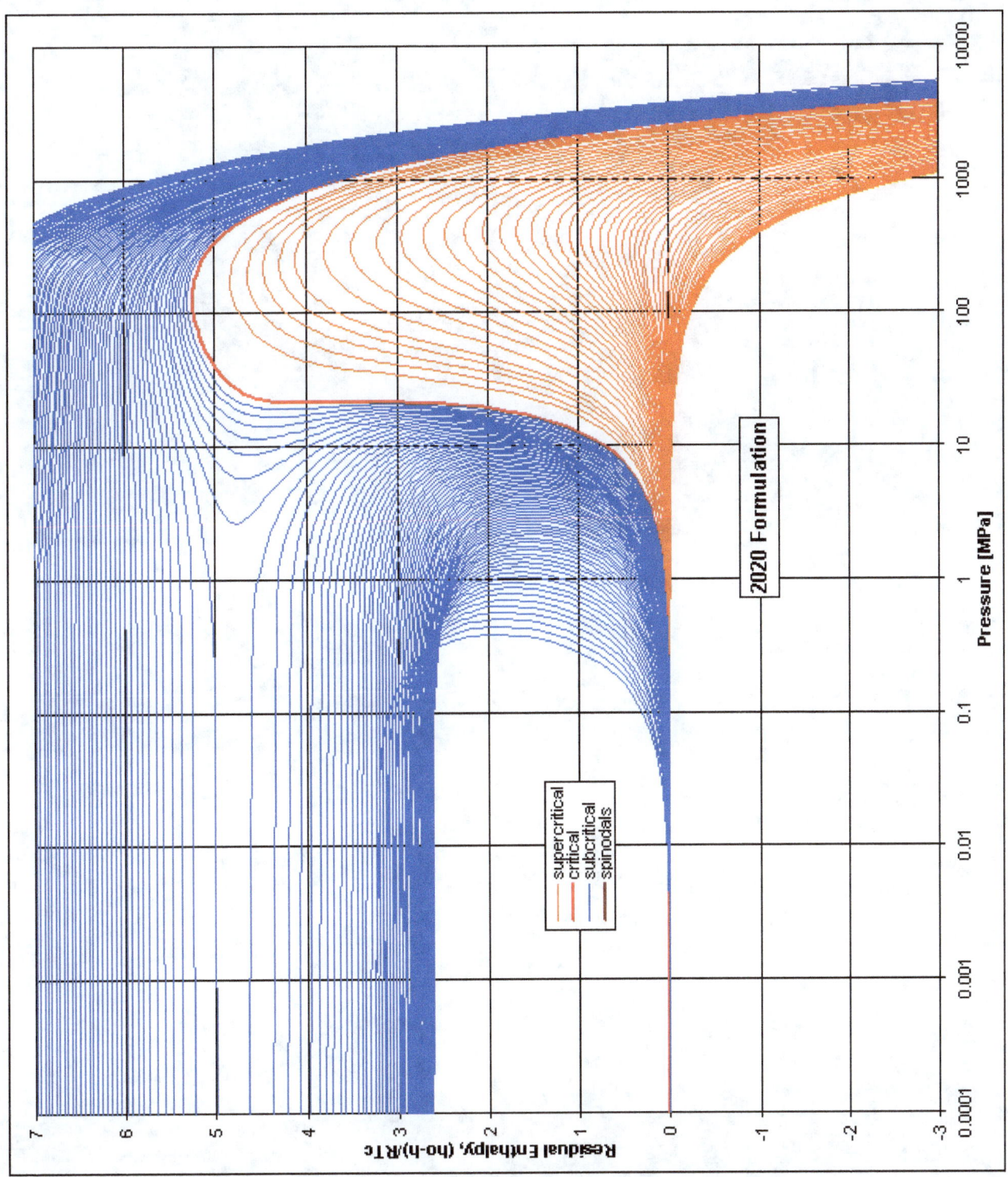

Figure 84. Residual Enthalpy Based on Steam 2020 Formulation

Note that this figure also shows the metastable regions, which the others do not.

Chapter 7. Residual Entropy

Residual entropy is like residual enthalpy in that it is the departure from ideal behavior and can be expressed by the following:

$$\frac{S_0 - S}{R} = -\ln Z + \int_{\infty}^{V_R} \left(\frac{1}{V_R} - \frac{\partial P_R}{\partial V_R} \right) dV_R \qquad (7.1)$$

Again, S_0 is the ideal gas entropy and S is the real fluid entropy. Through a convoluted process, one can show that the residual entropy is also related to the fugacity coefficient:

$$\frac{S_0 - S}{R} = \frac{H_0 - H}{RT} + \ln F \qquad (7.2)$$

In terms of reduced quantities (more compact notation):

$$S_R = \frac{H_R}{T_R} + \ln F \qquad (7.3)$$

The residual entropy for the van der Waals EOS is:

$$S_R = \ln\left(\frac{V_R}{V_R - B}\right) - \ln Z \qquad (7.4)$$

The remaining expressions are calculated using Equation 7.3.

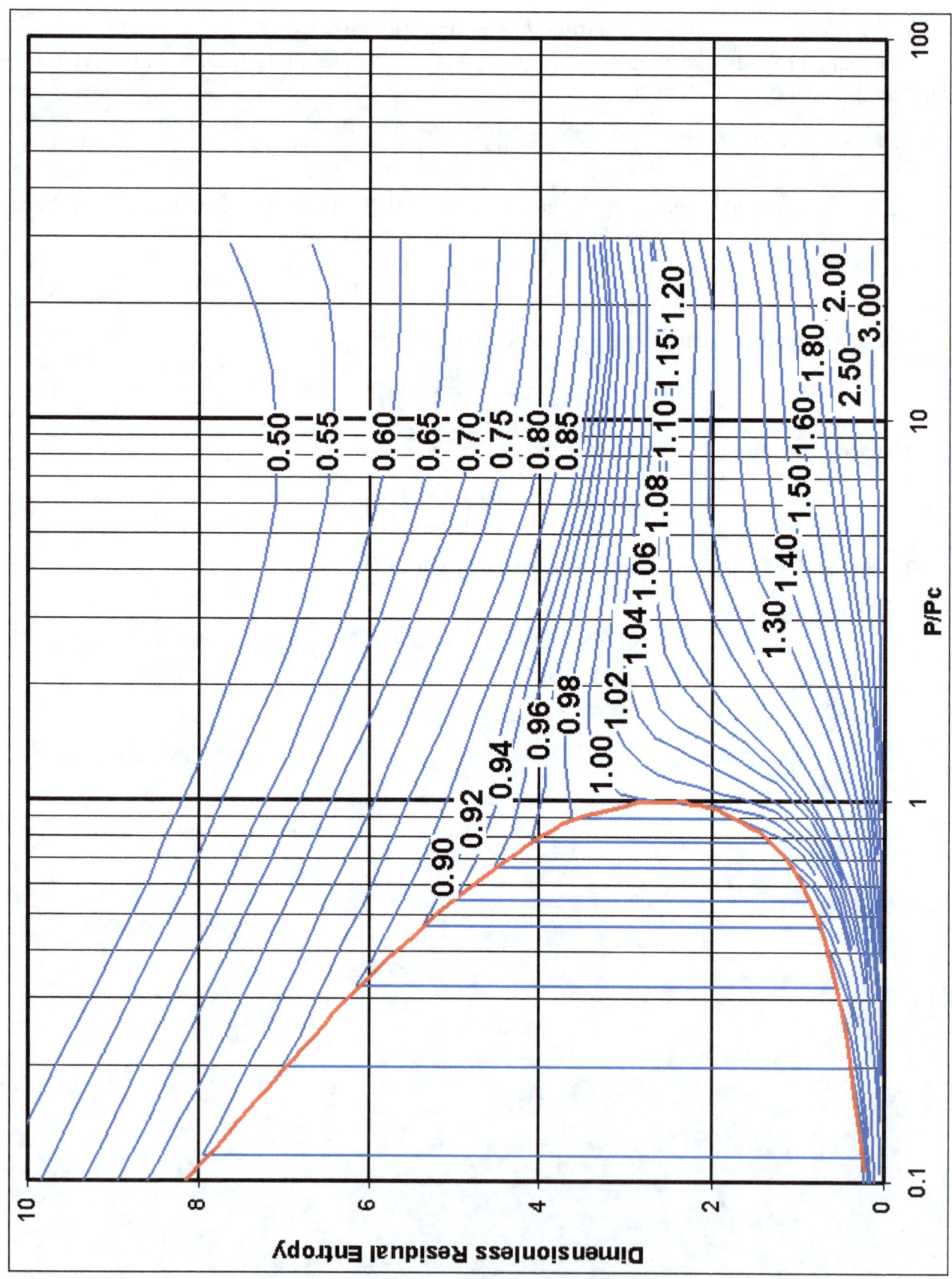

Figure 85. Residual Entropy Based on Nelson-Obert

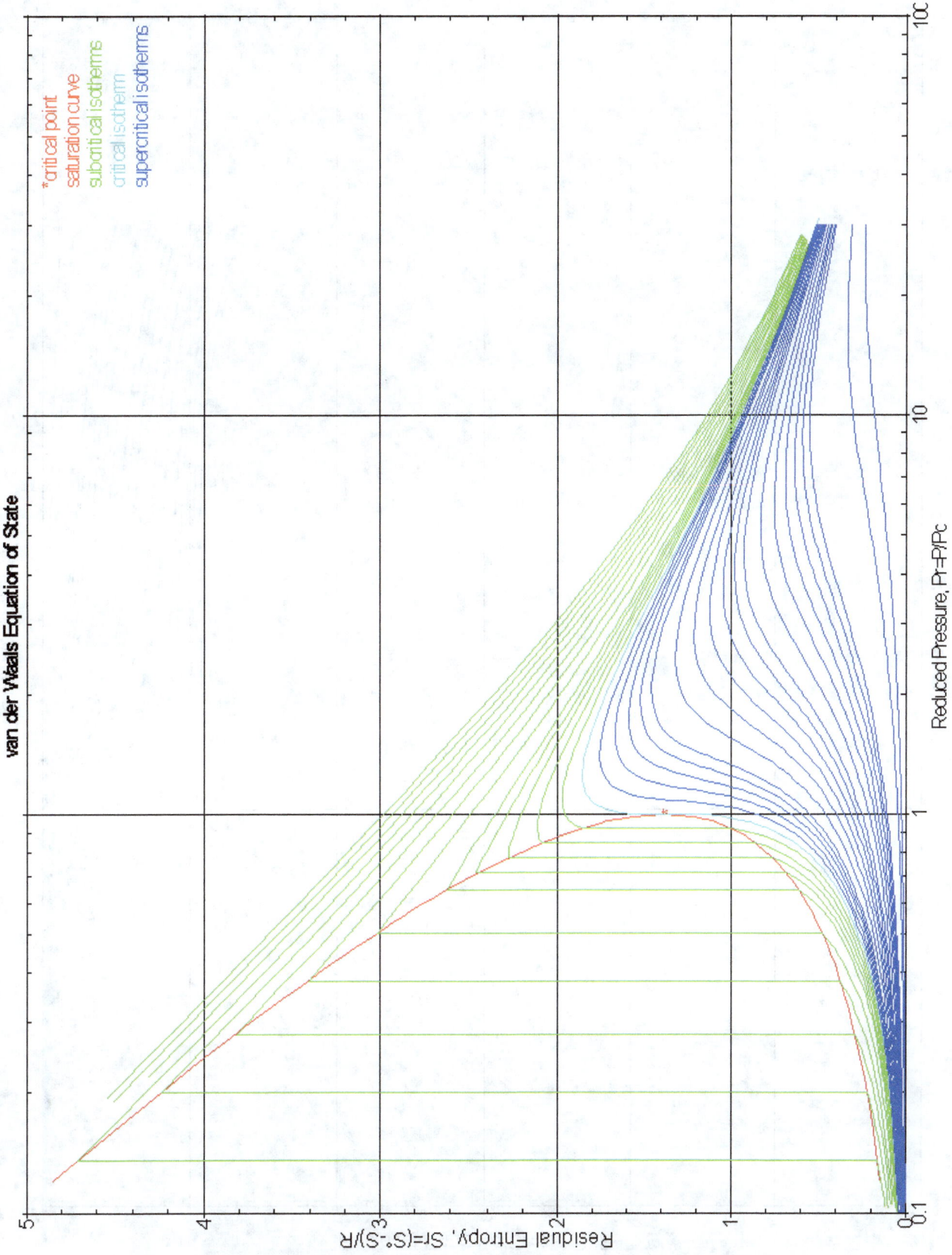

Figure 86. Residual Entropy Based on van der Waals

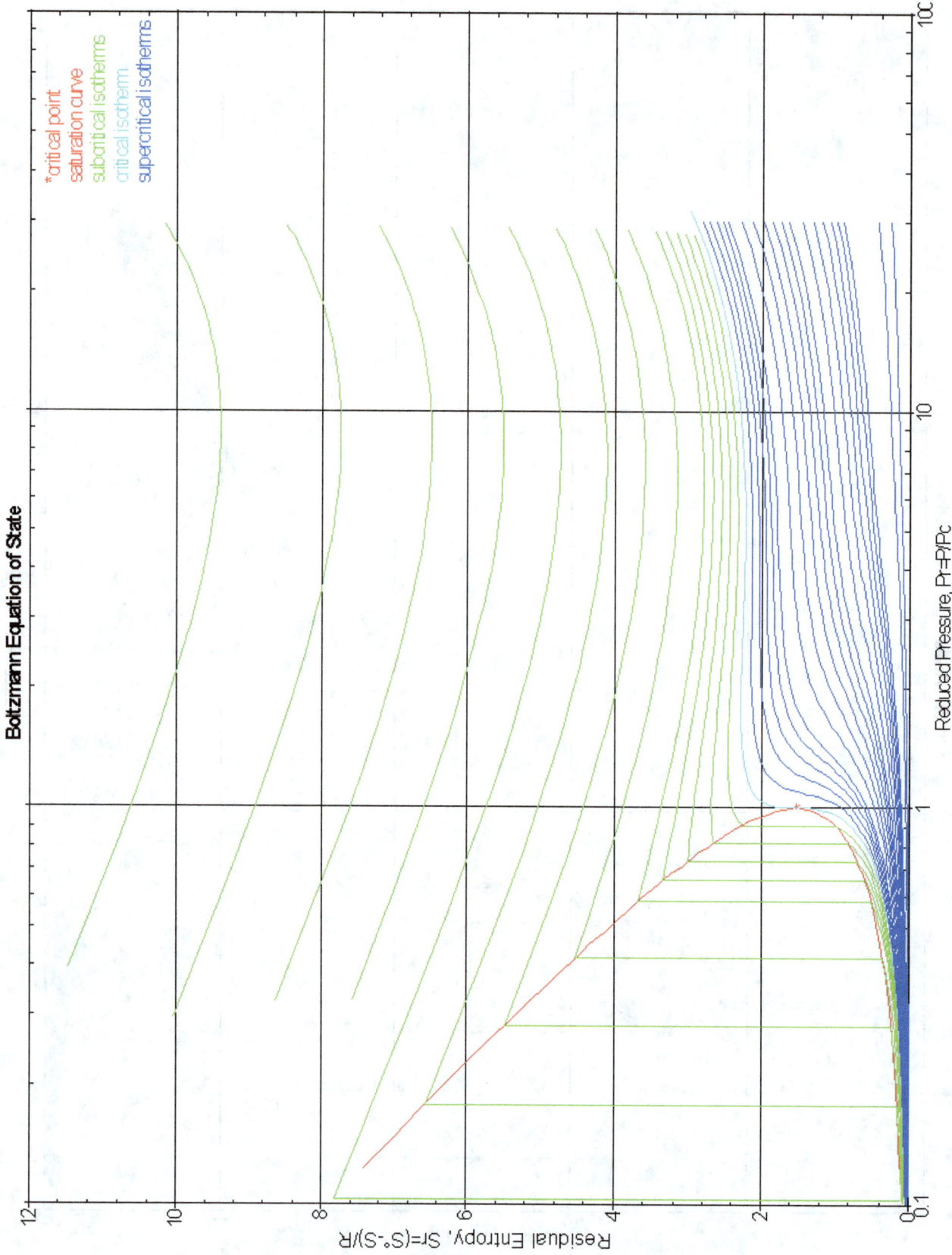

Figure 87. Residual Entropy Based on Boltzmann

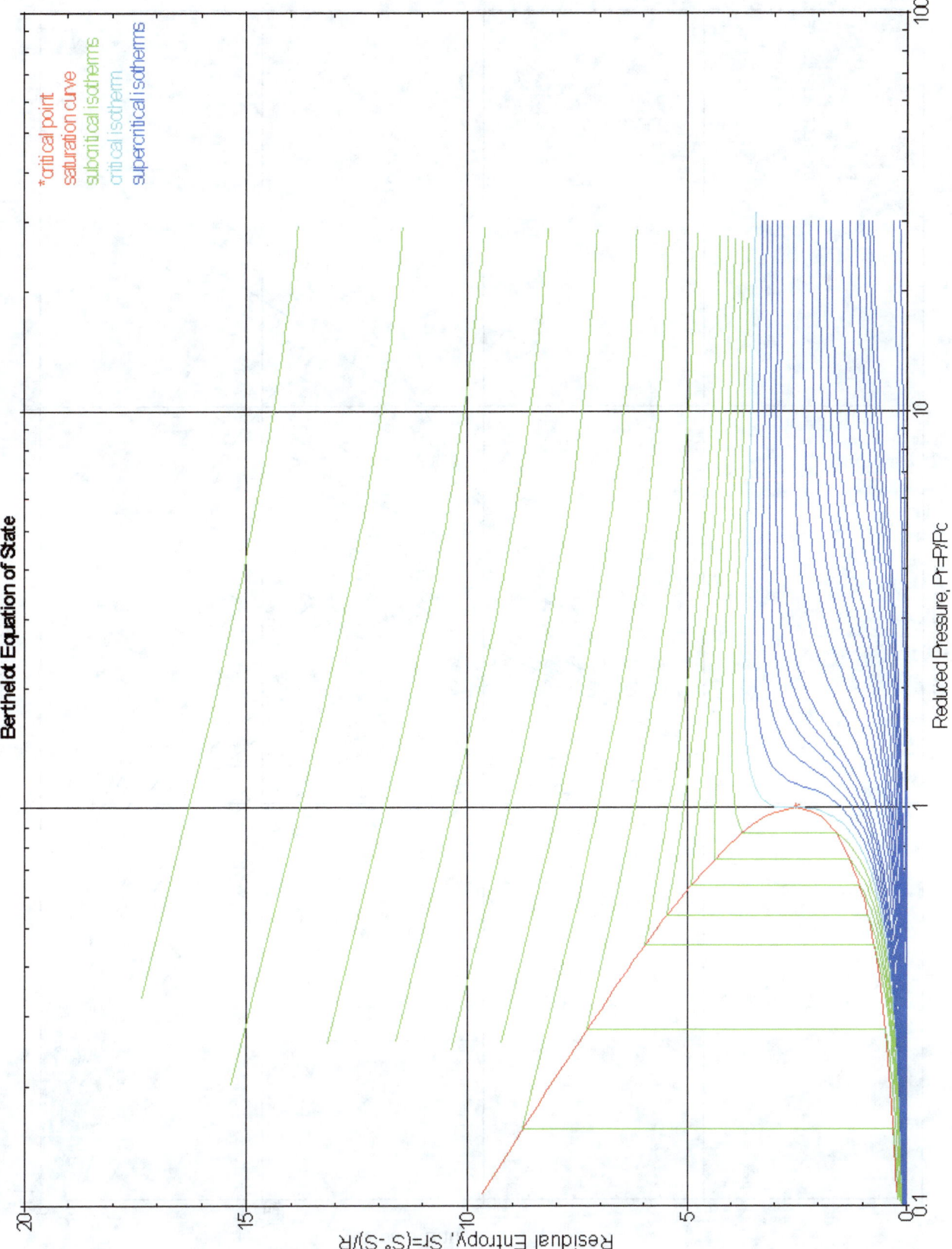

Figure 88. Residual Entropy Based on Berthelot

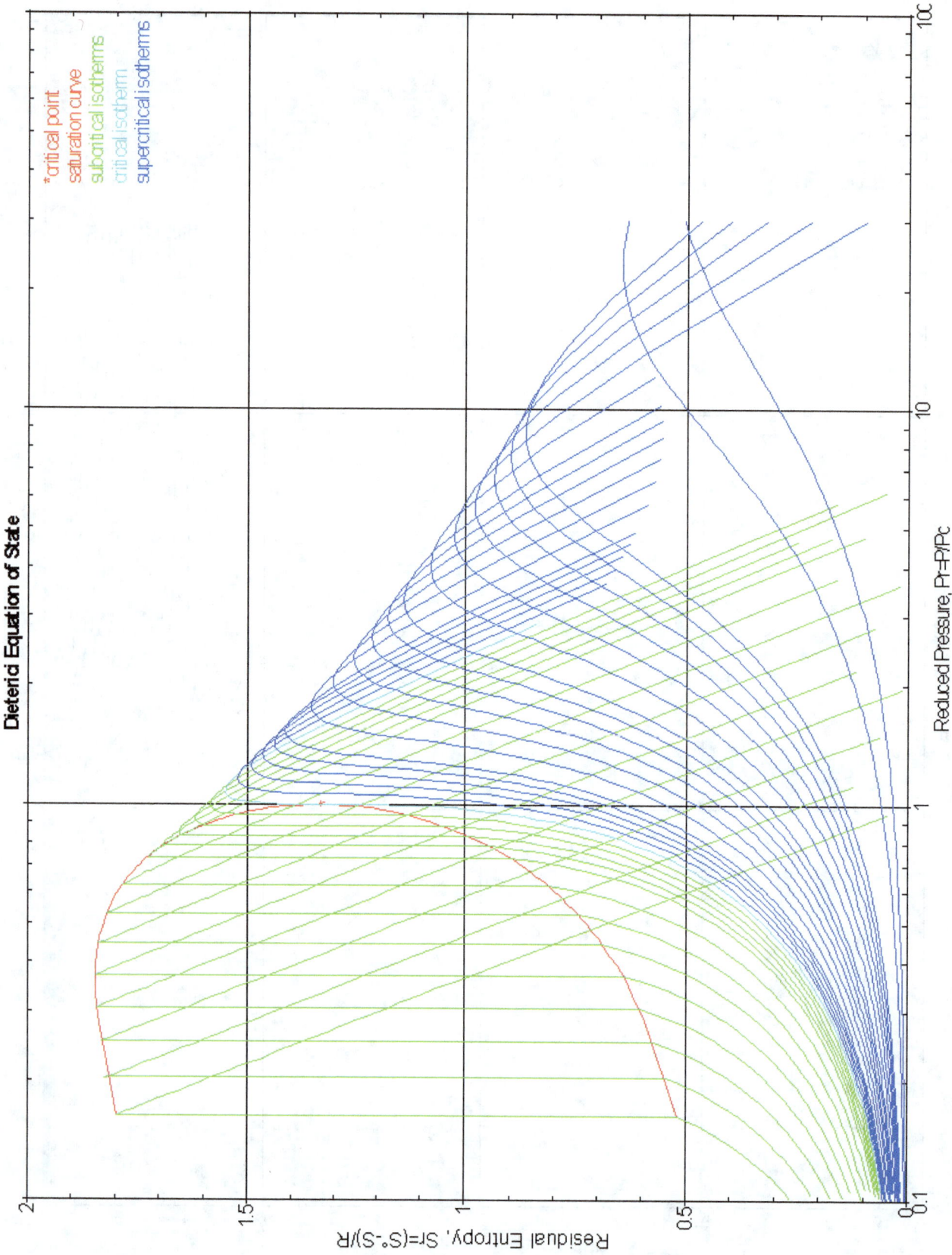

Figure 89. Residual Entropy Based on Dieterici

The behavior illustrated in the figure above is unacceptable.

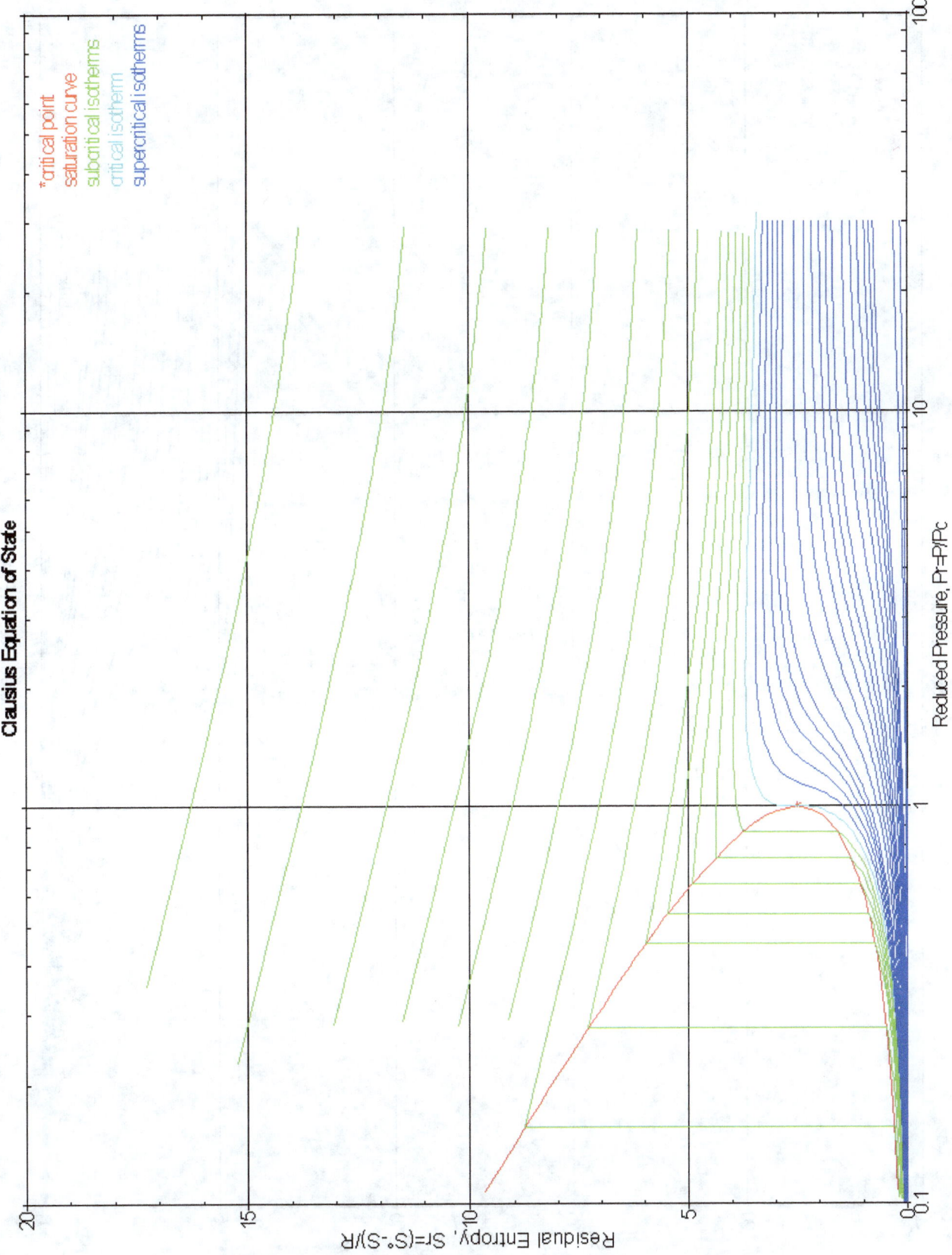

Figure 90. Residual Entropy Based on Clausius

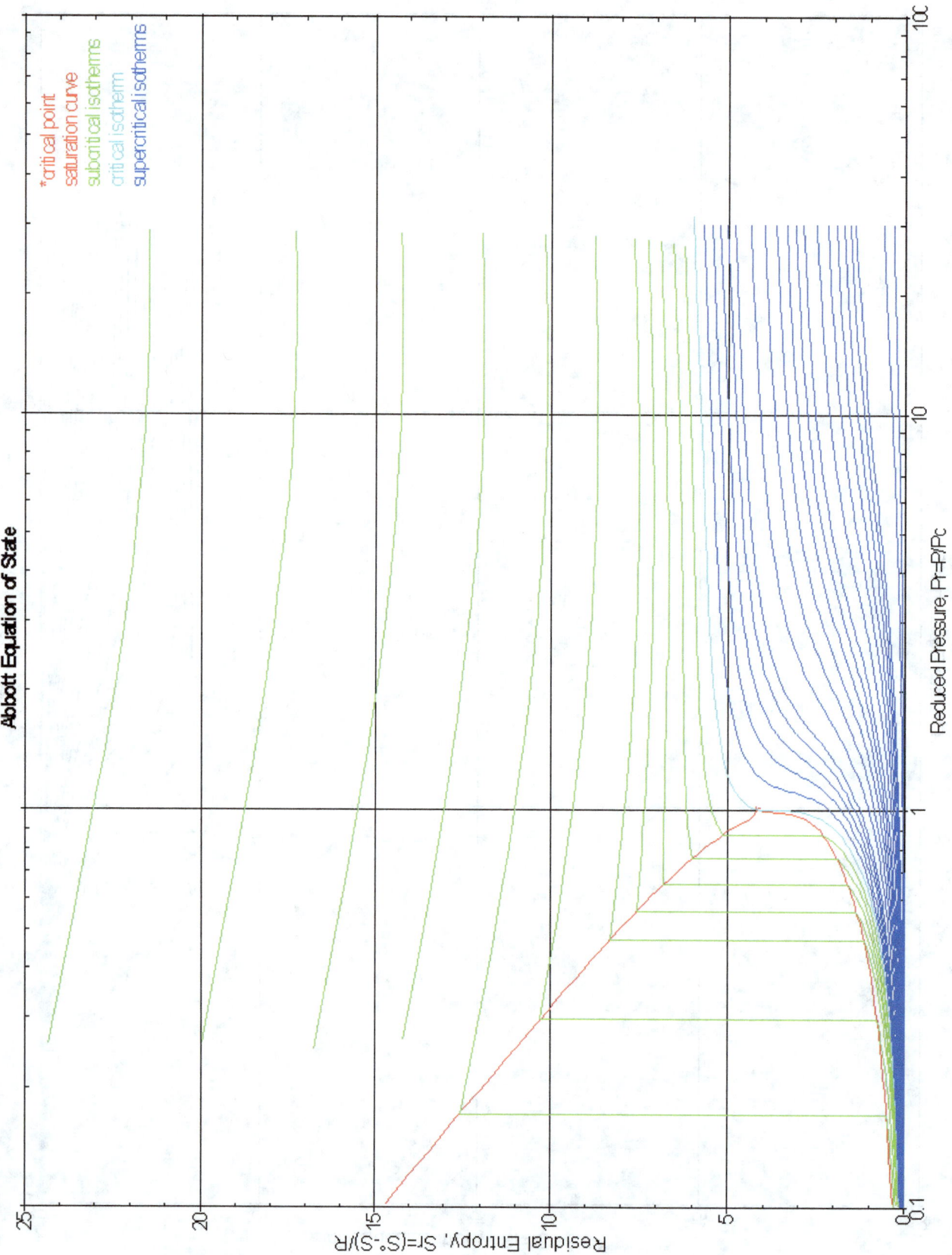

Figure 91. Residual Entropy Based on Abbott's Modification

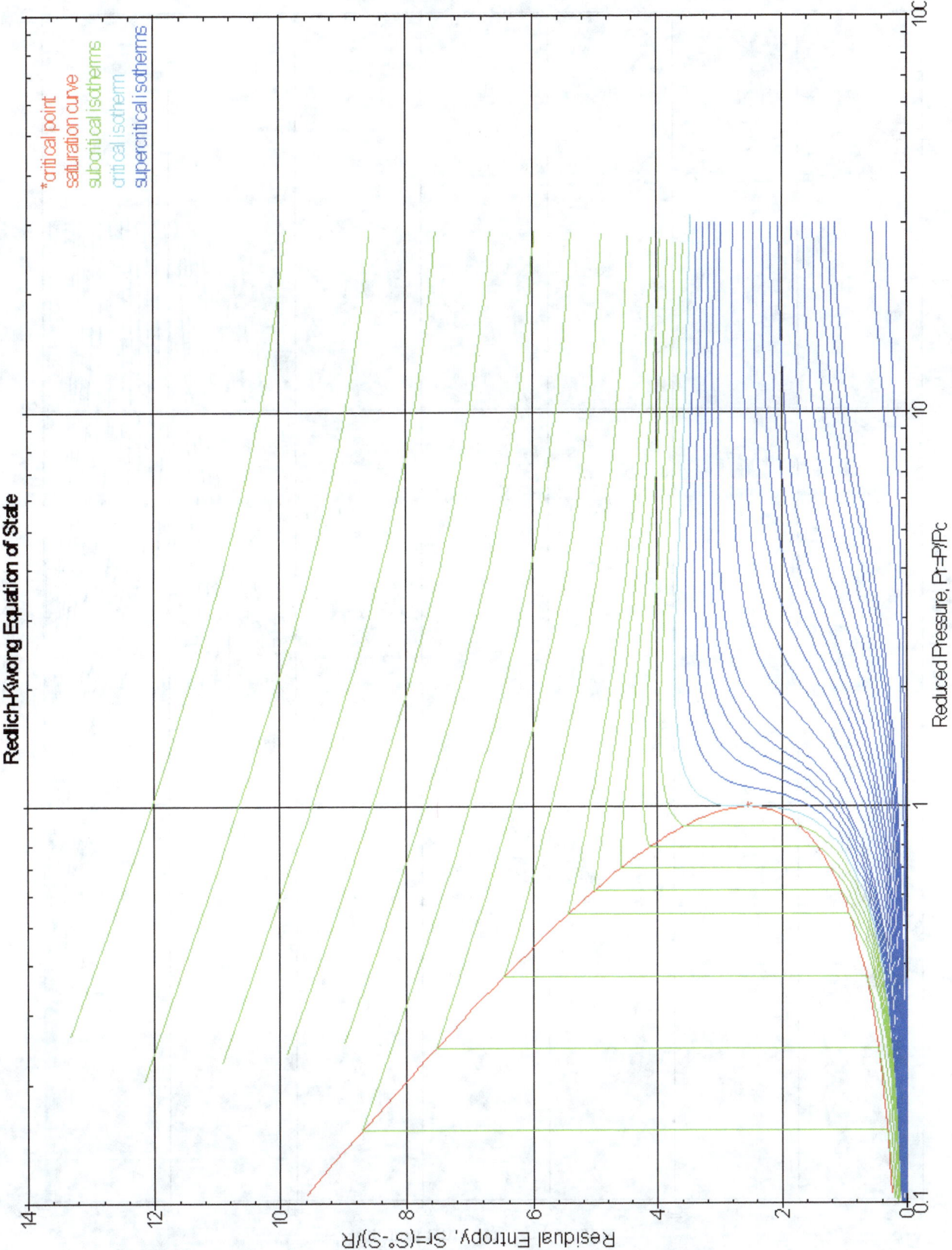

Figure 92. Residual Entropy Based on Redlich-Kwong

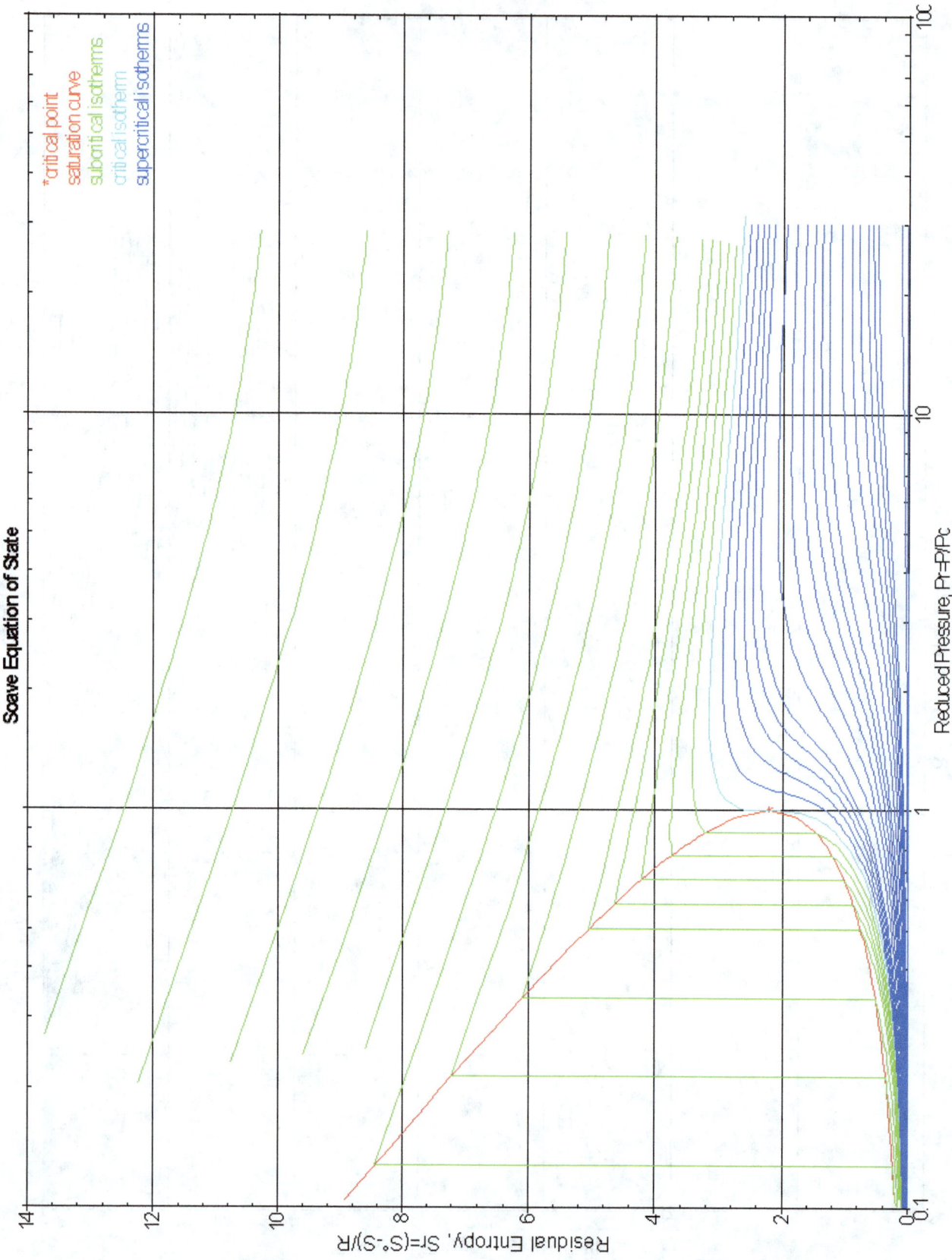

Figure 93. Residual Entropy Based on Soave's Modification

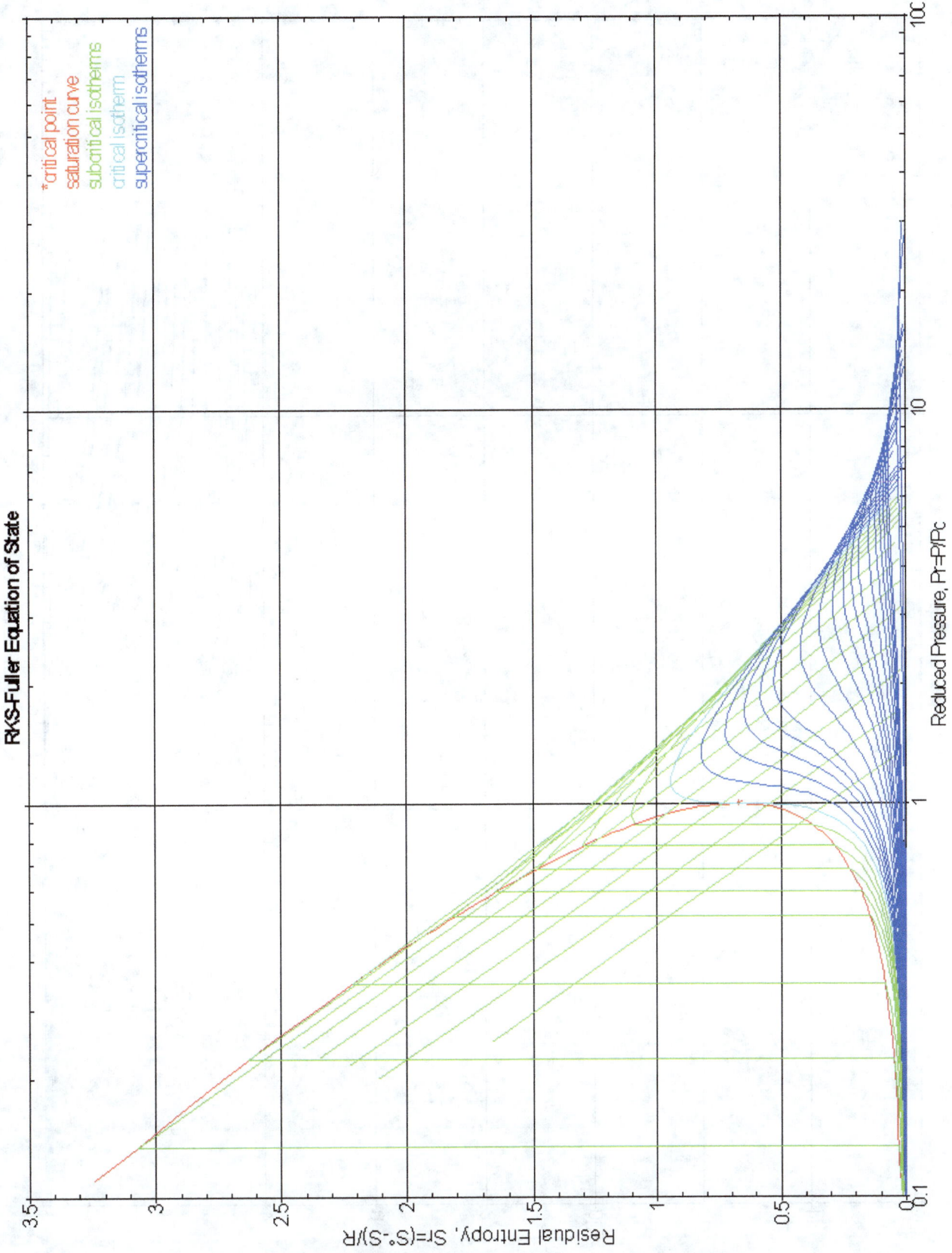

Figure 94. Residual Entropy Based on Fuller's Modification

The behavior illustrated in the figure above is unacceptable.

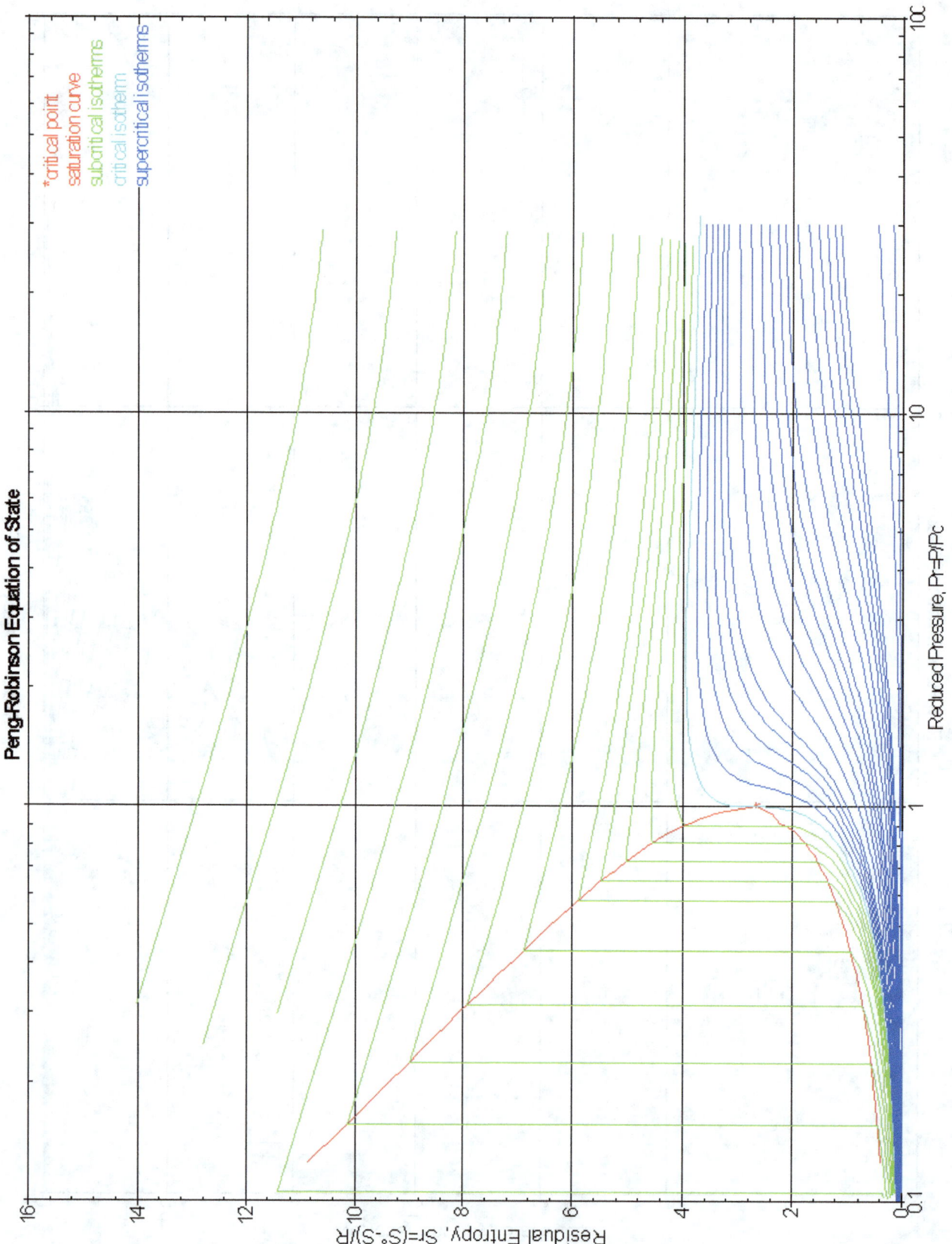

Figure 95. Residual Entropy Based on Peng-Robinson

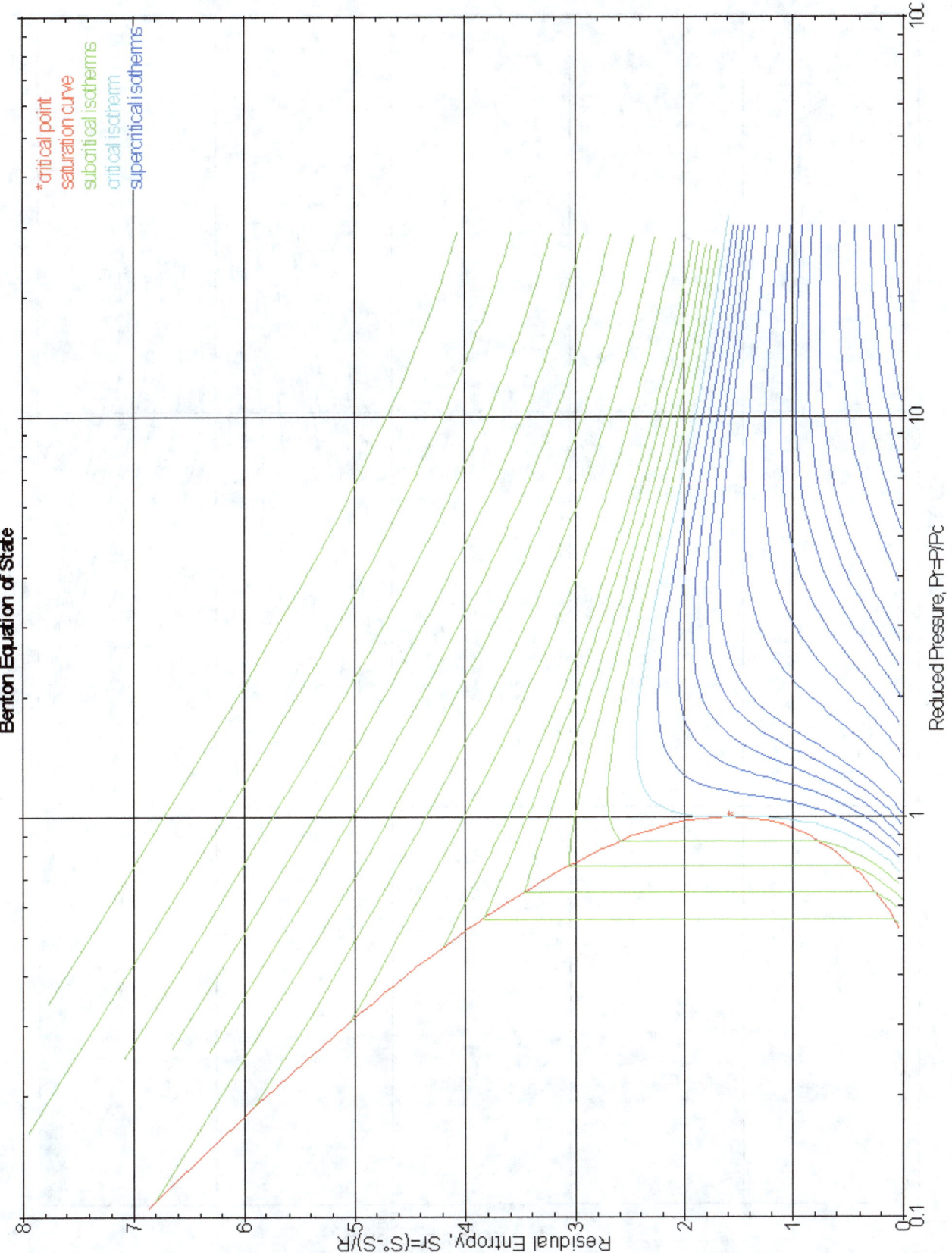

Figure 96. Residual Entropy Based on Author's Modification

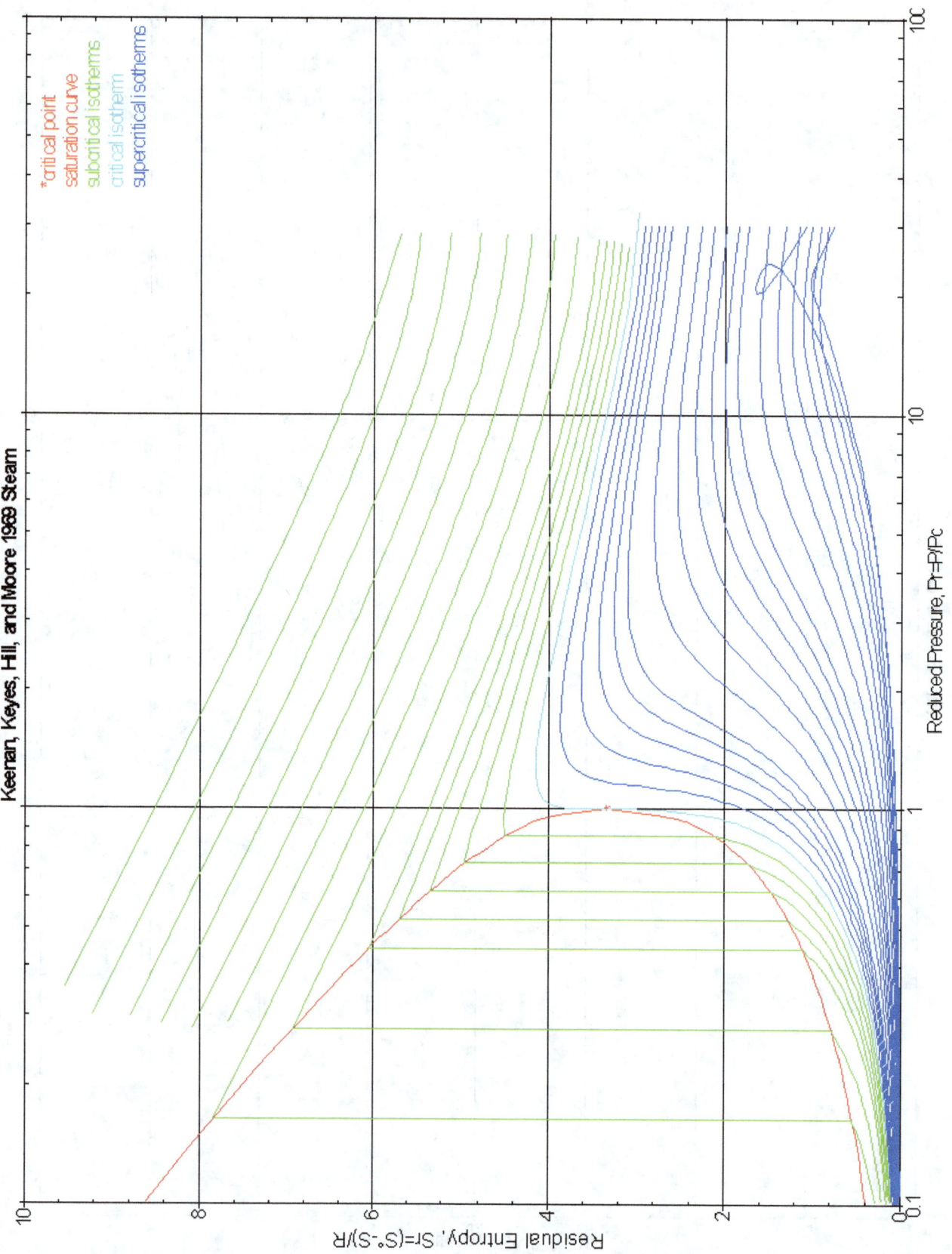

Figure 97. Residual Entropy Based on Keenan, Keyes, Hill, and Moore

The behavior at very high pressures shown above is wrong but beyond the recommended pressure limit for this formulation.

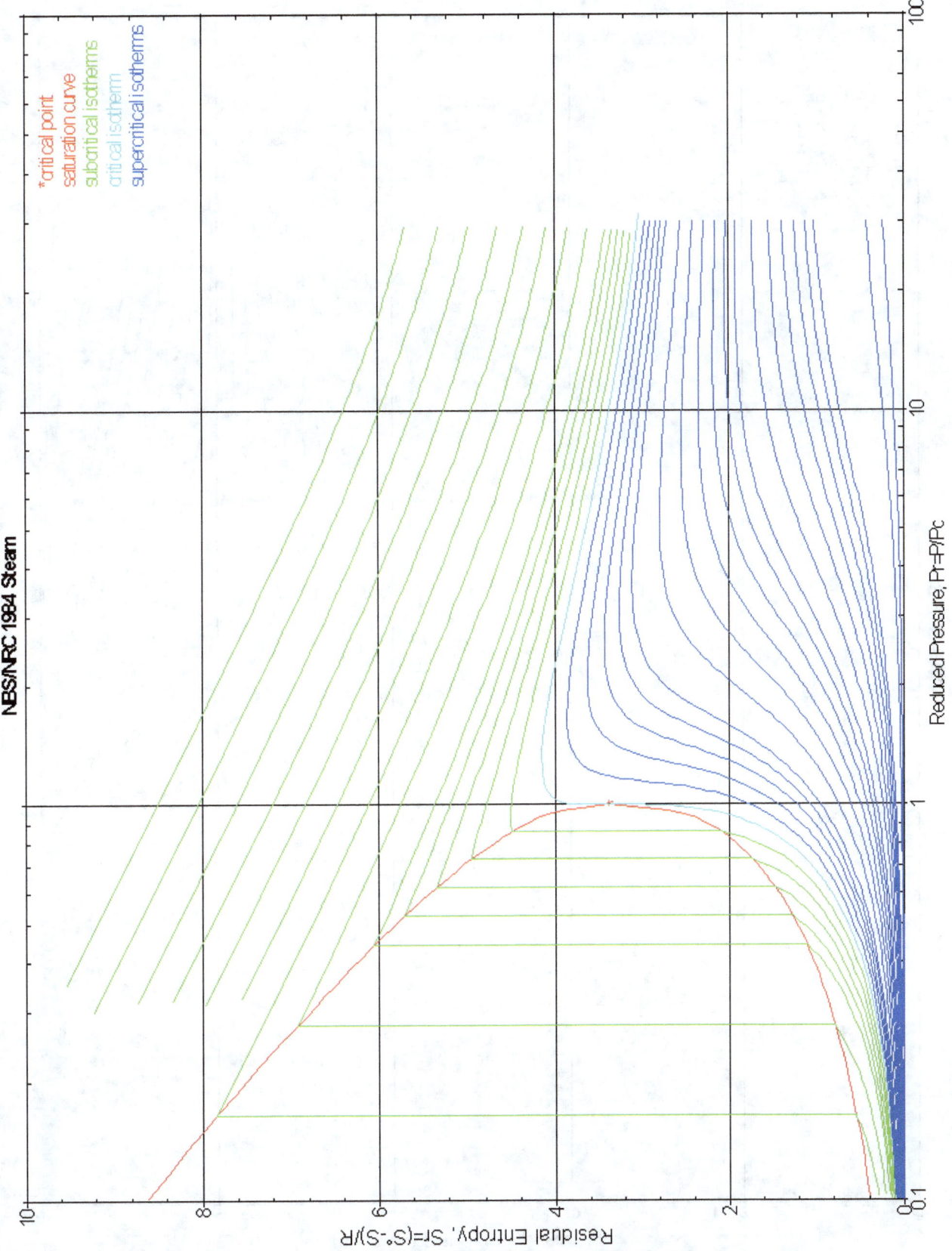

Figure 98. Residual Entropy Based on Haar, Gallagher, and Kell

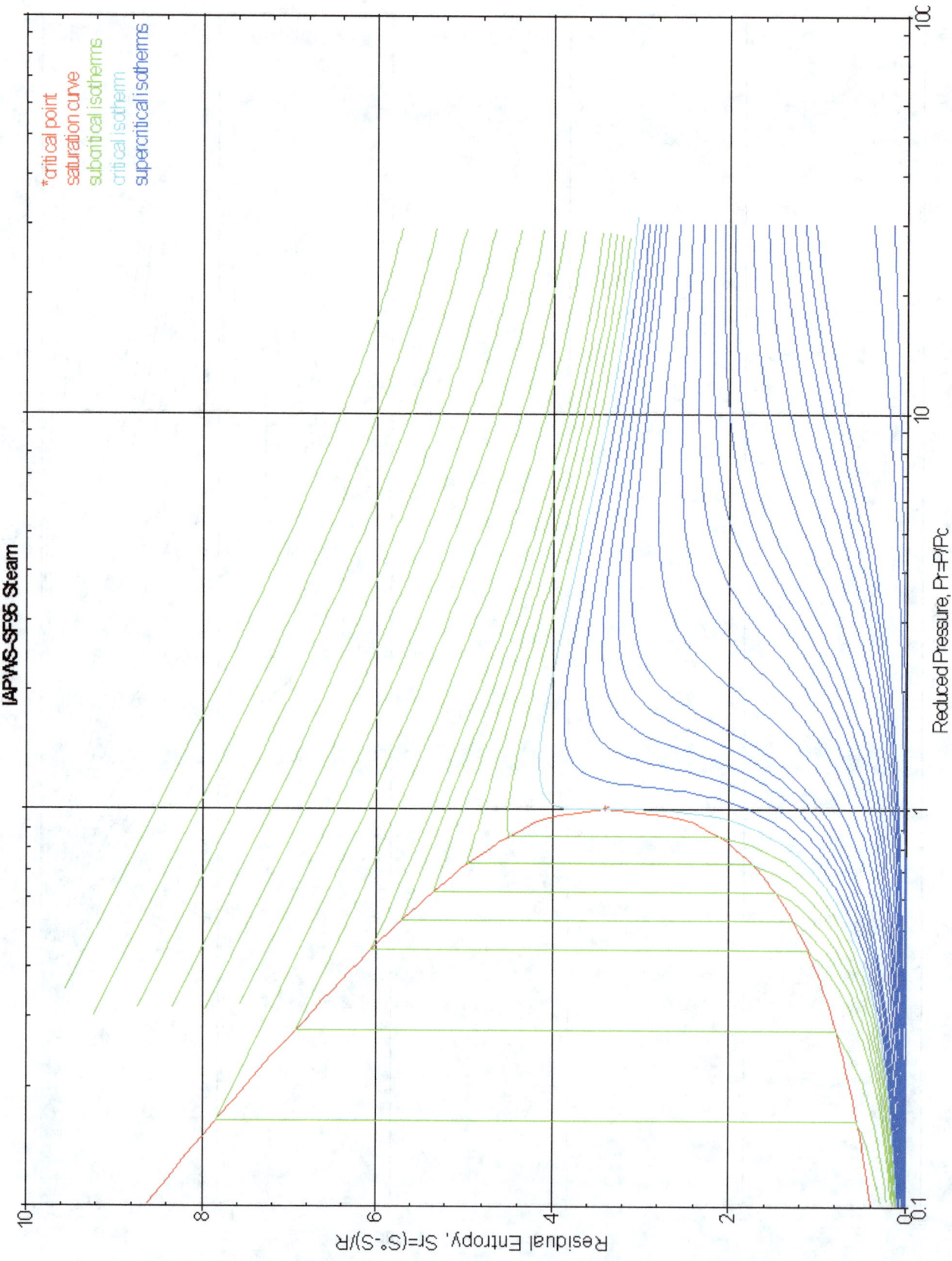

Figure 99. Residual Entropy Based on Wagner and Pruß

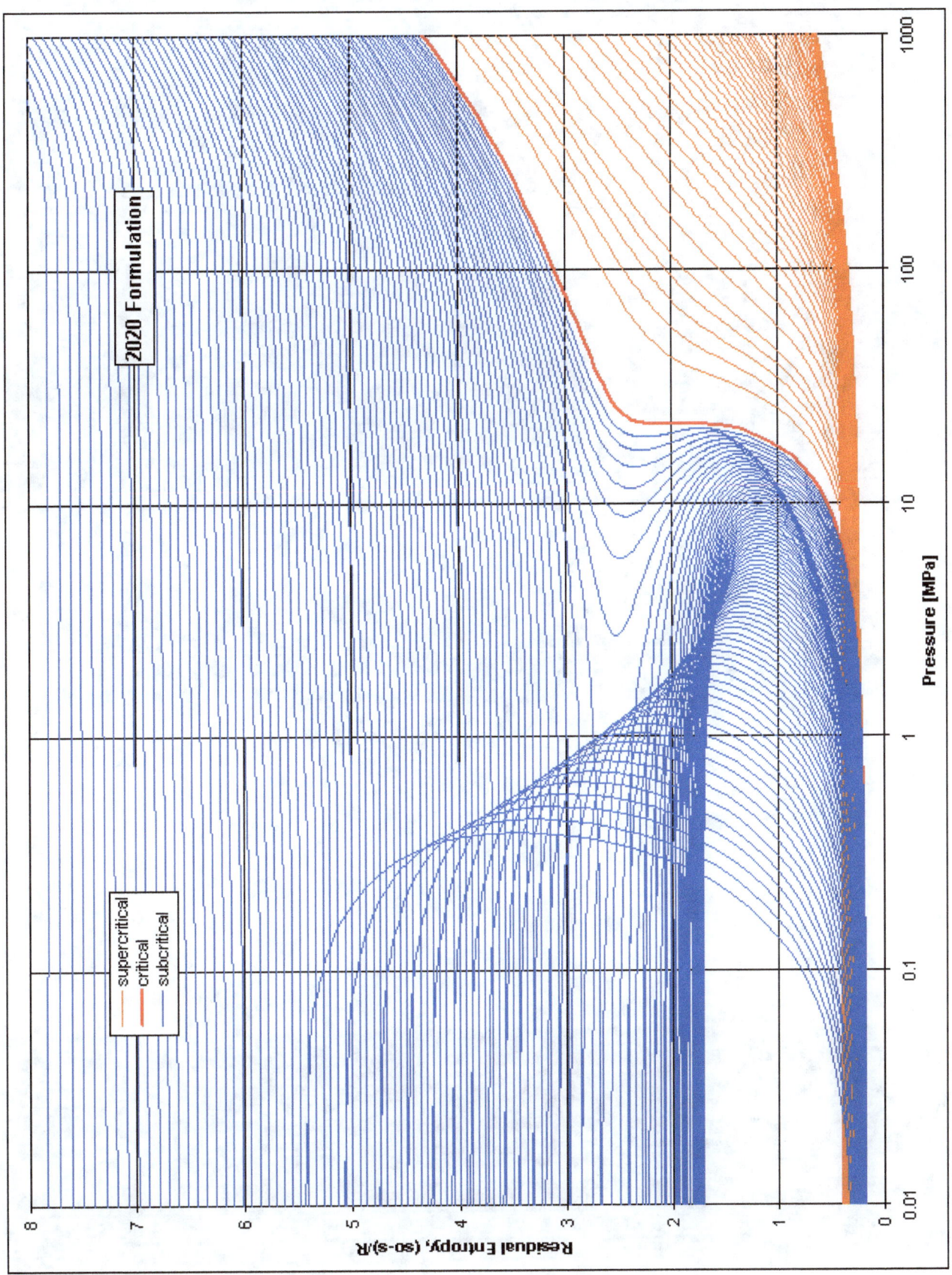

Figure 100. Residual Entropy Based on Steam 2020 Formulation

Note that this figure also shows the metastable regions, which the others do not.

110

Chapter 8. Pressure vs. Enthalpy

The P-H chart is often used in analysis of refrigeration systems and so charts of this type are often found in references such as the ASHRAE Handbook of Fundamentals. As we are not introducing any new formula in this chapter, we begin with the Nelson-Obert Data.

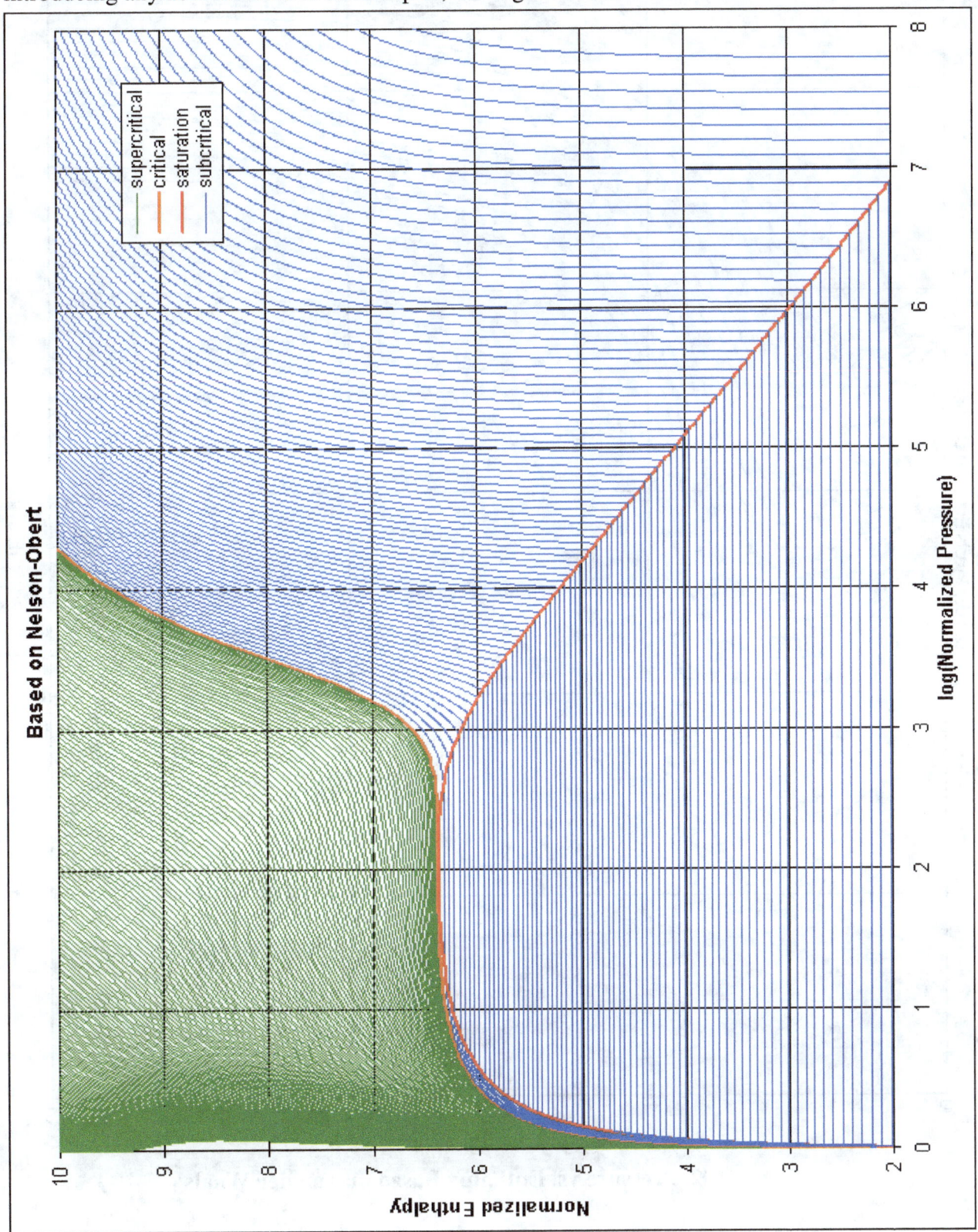

Figure 101. Pressure vs. Enthalpy Based on Nelson-Obert

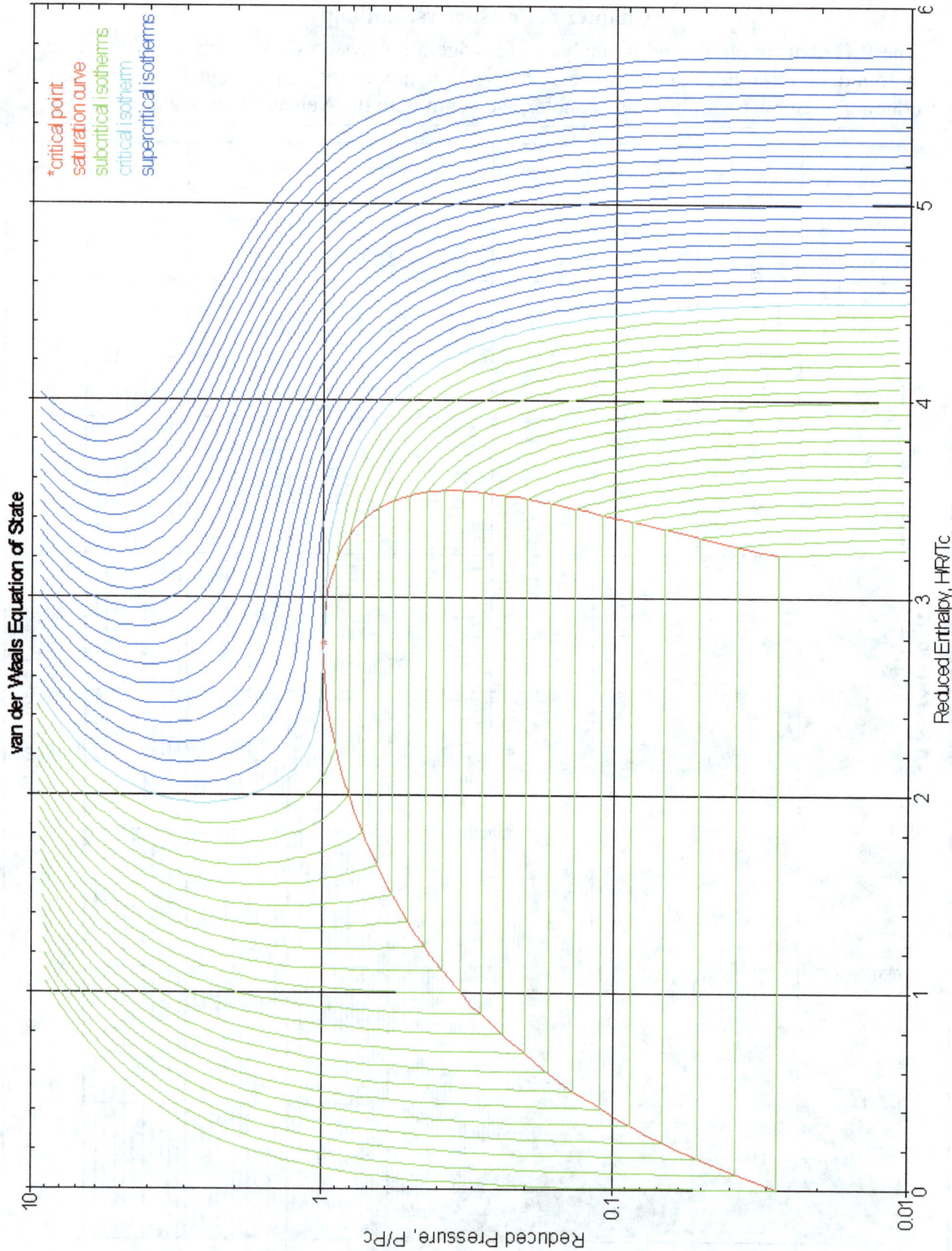

Figure 102. Pressure vs. Enthalpy Based on van der Waals

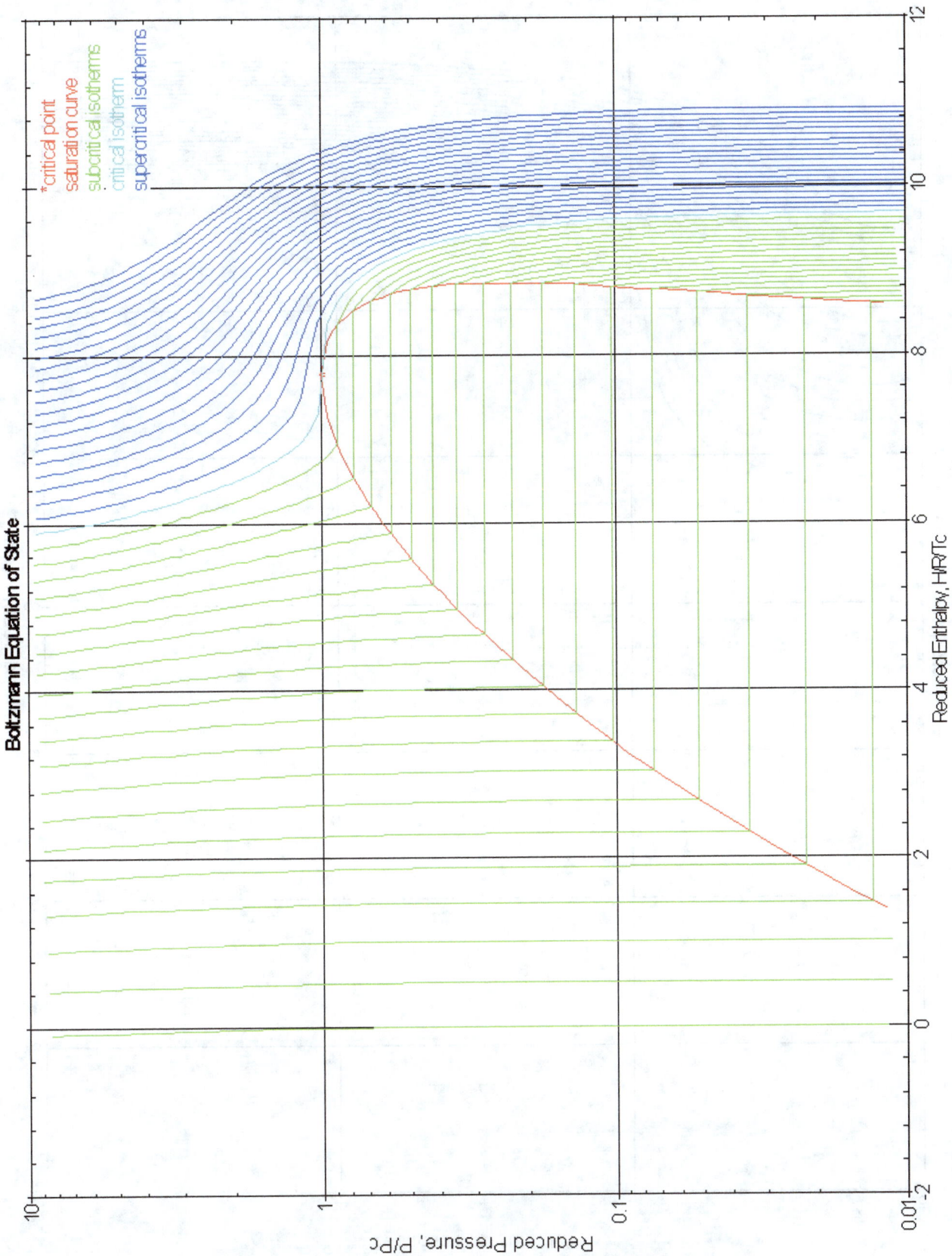

Figure 103. Pressure vs. Enthalpy Based on Boltzmann

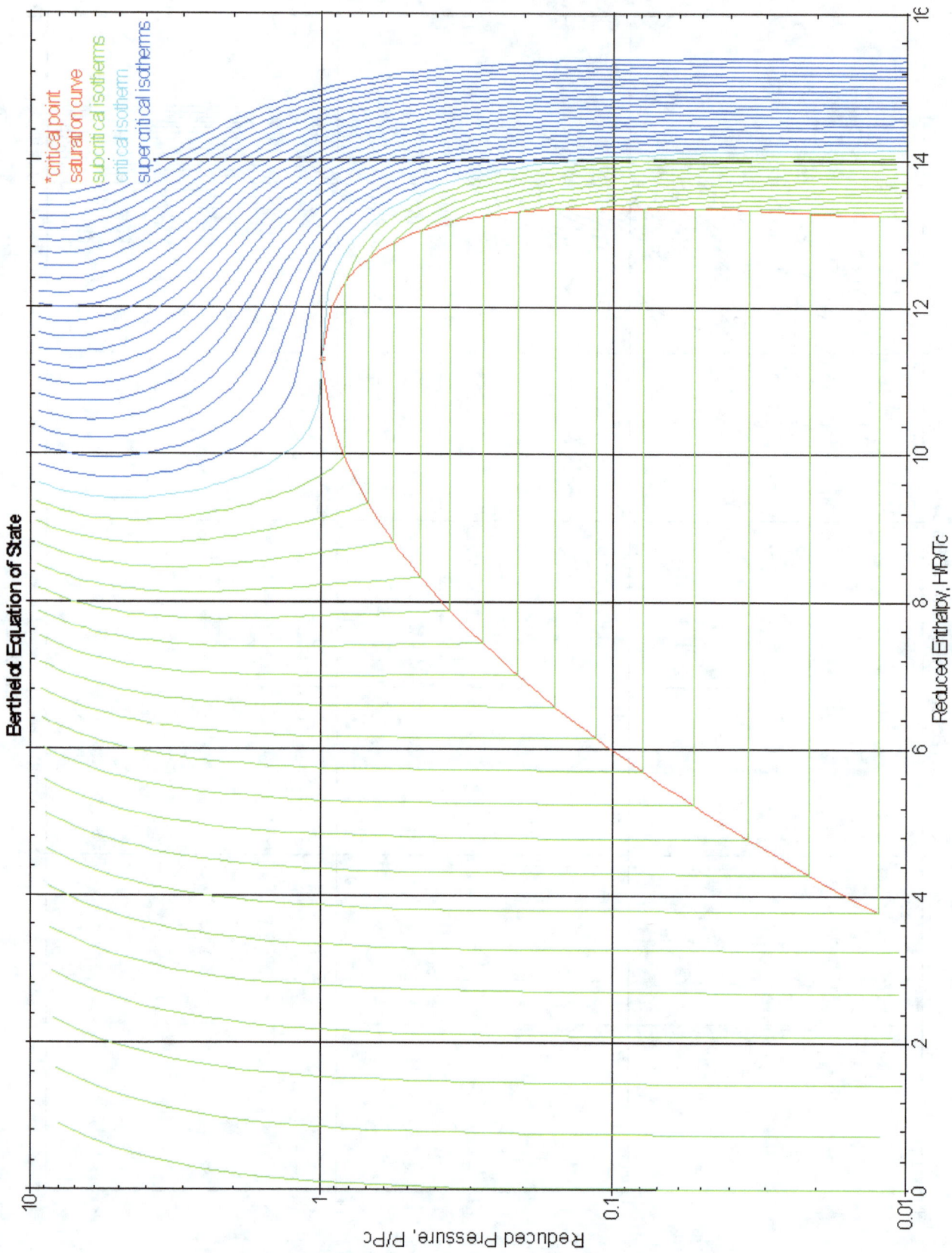

Figure 104. Pressure vs. Enthalpy Based on Berthelot

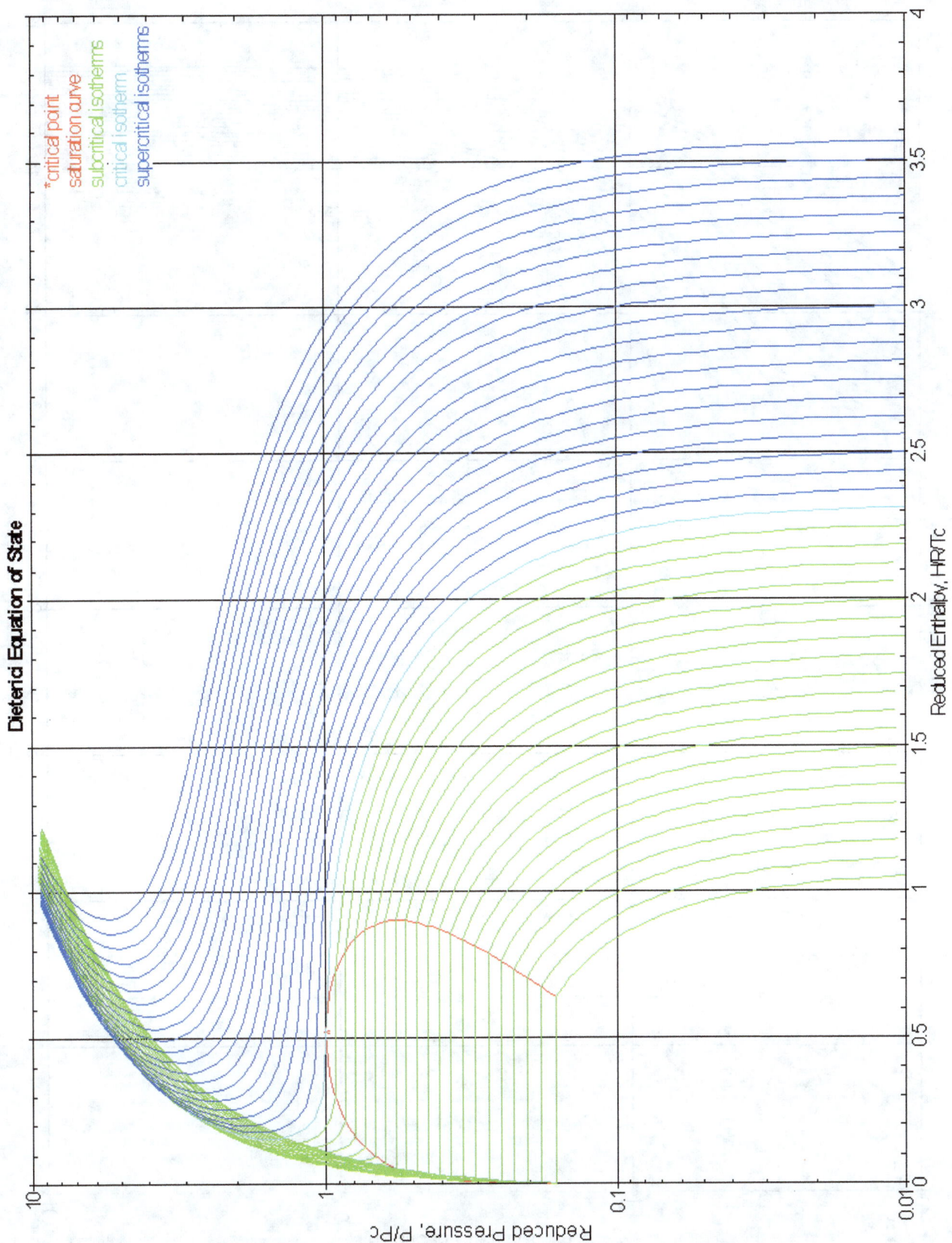

Figure 105. Pressure vs. Enthalpy Based on Dieterici

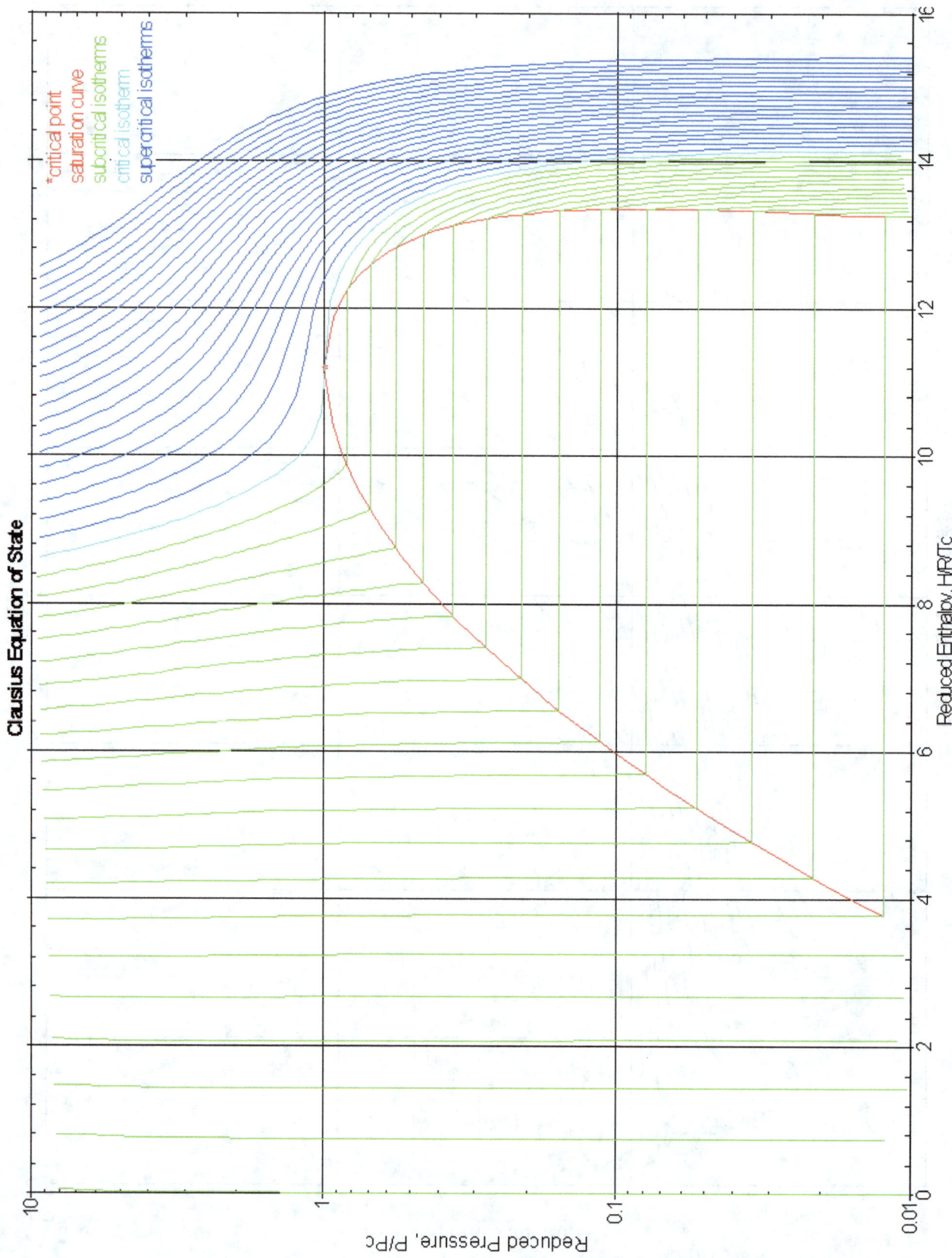

Figure 106. Pressure vs. Enthalpy Based on Clausius

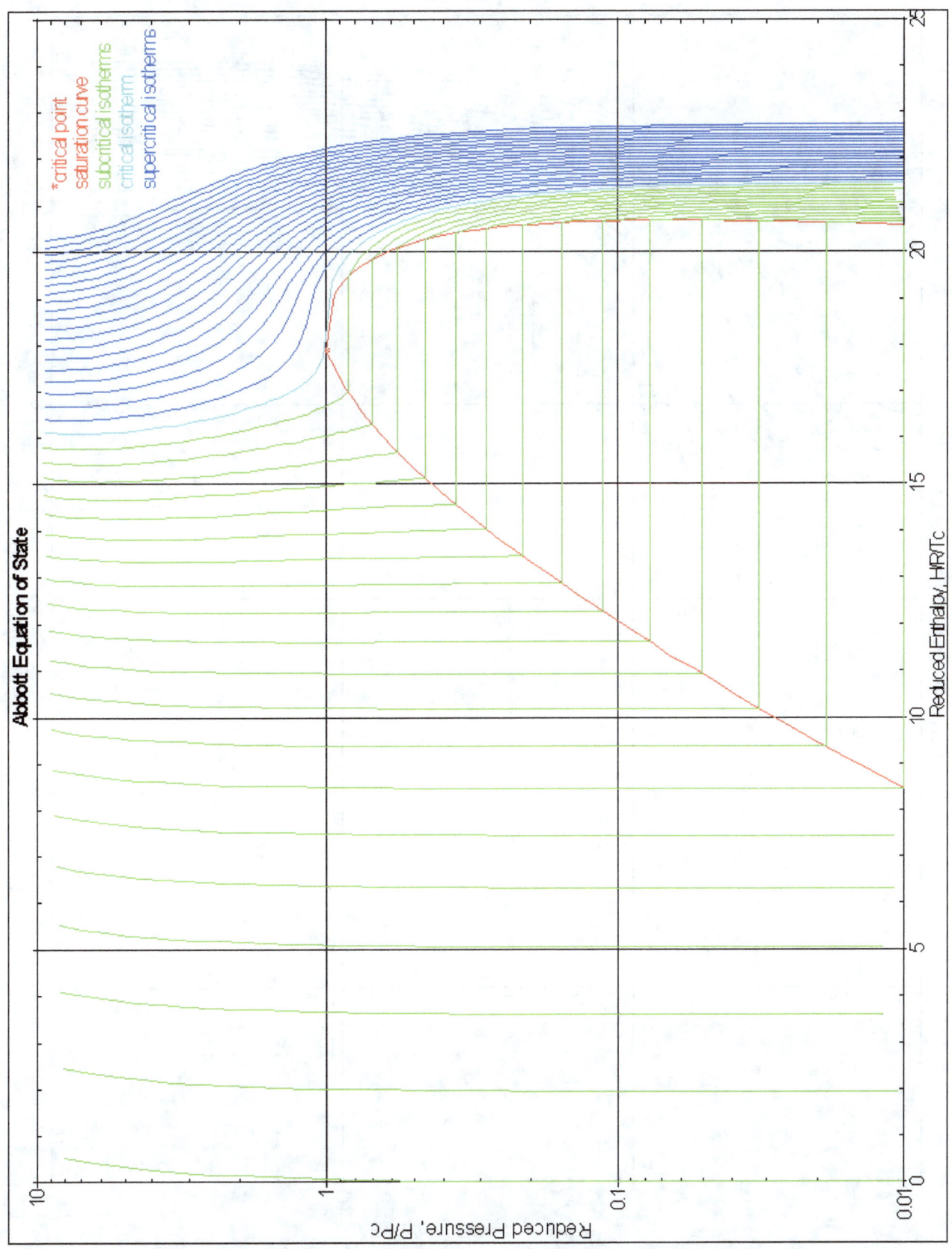

Figure 107. Pressure vs. Enthalpy Based on Abbott's Modification

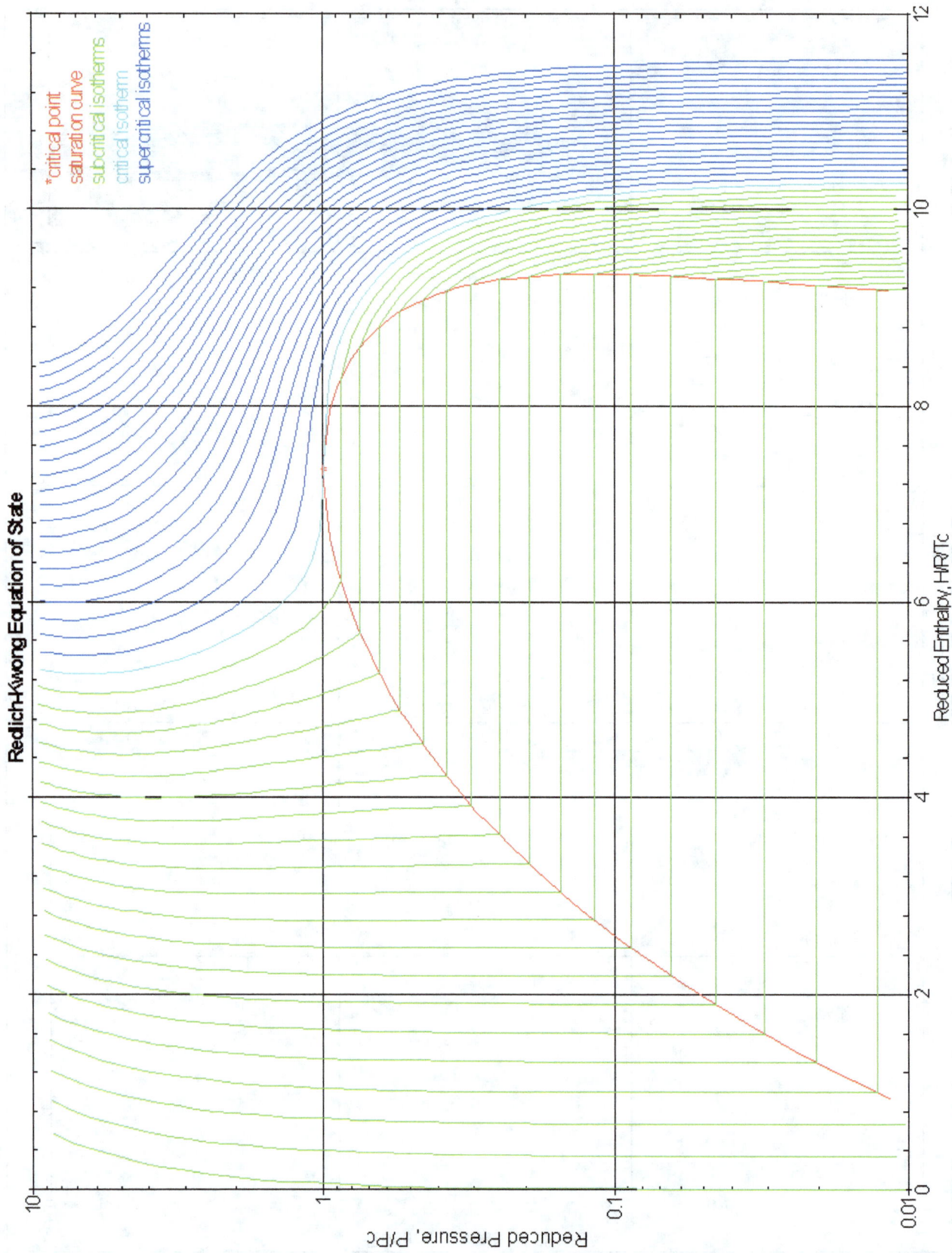

Figure 108. Pressure vs. Enthalpy Based on Redlich-Kwong

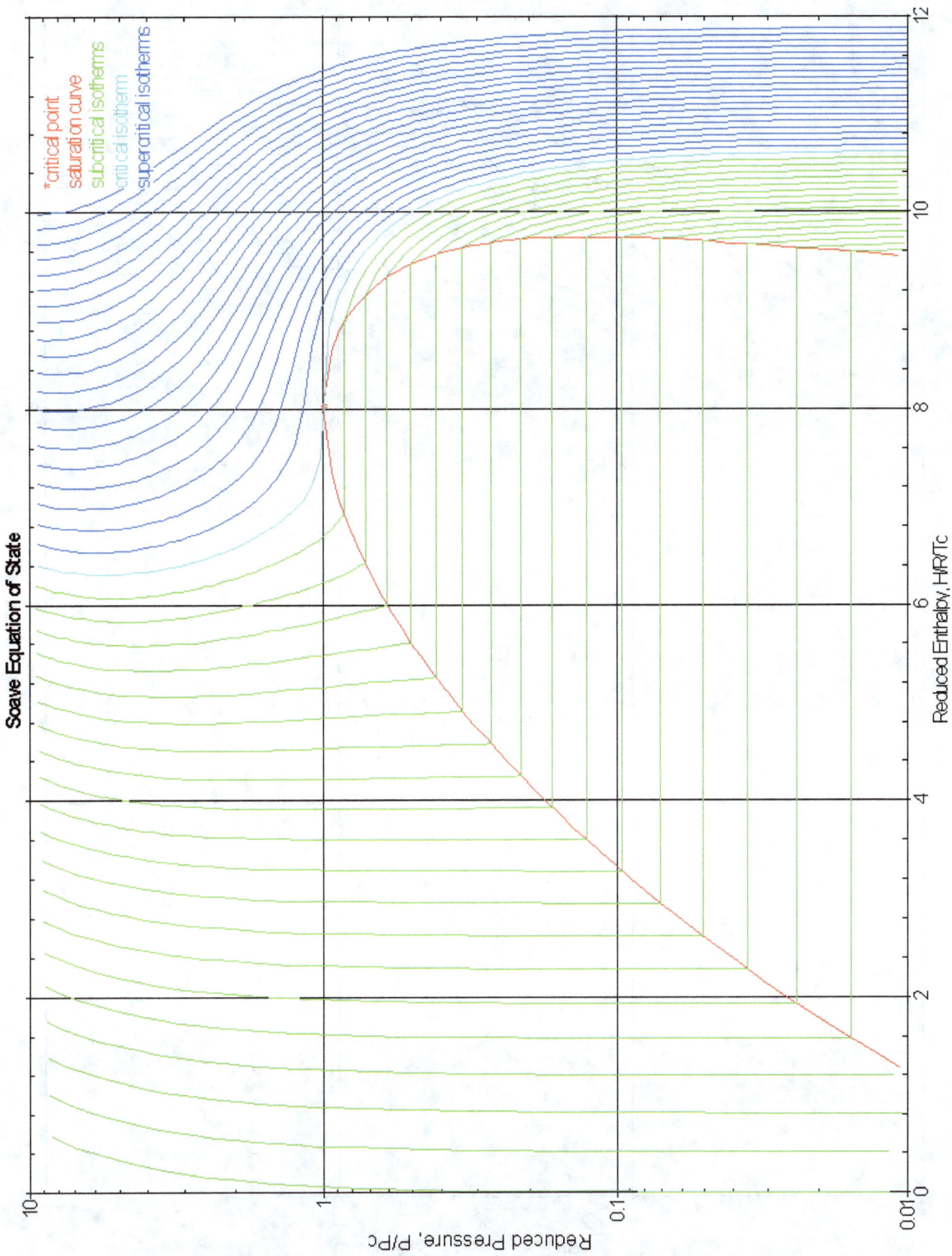

Figure 109. Pressure vs. Enthalpy Based on Soave's Modification

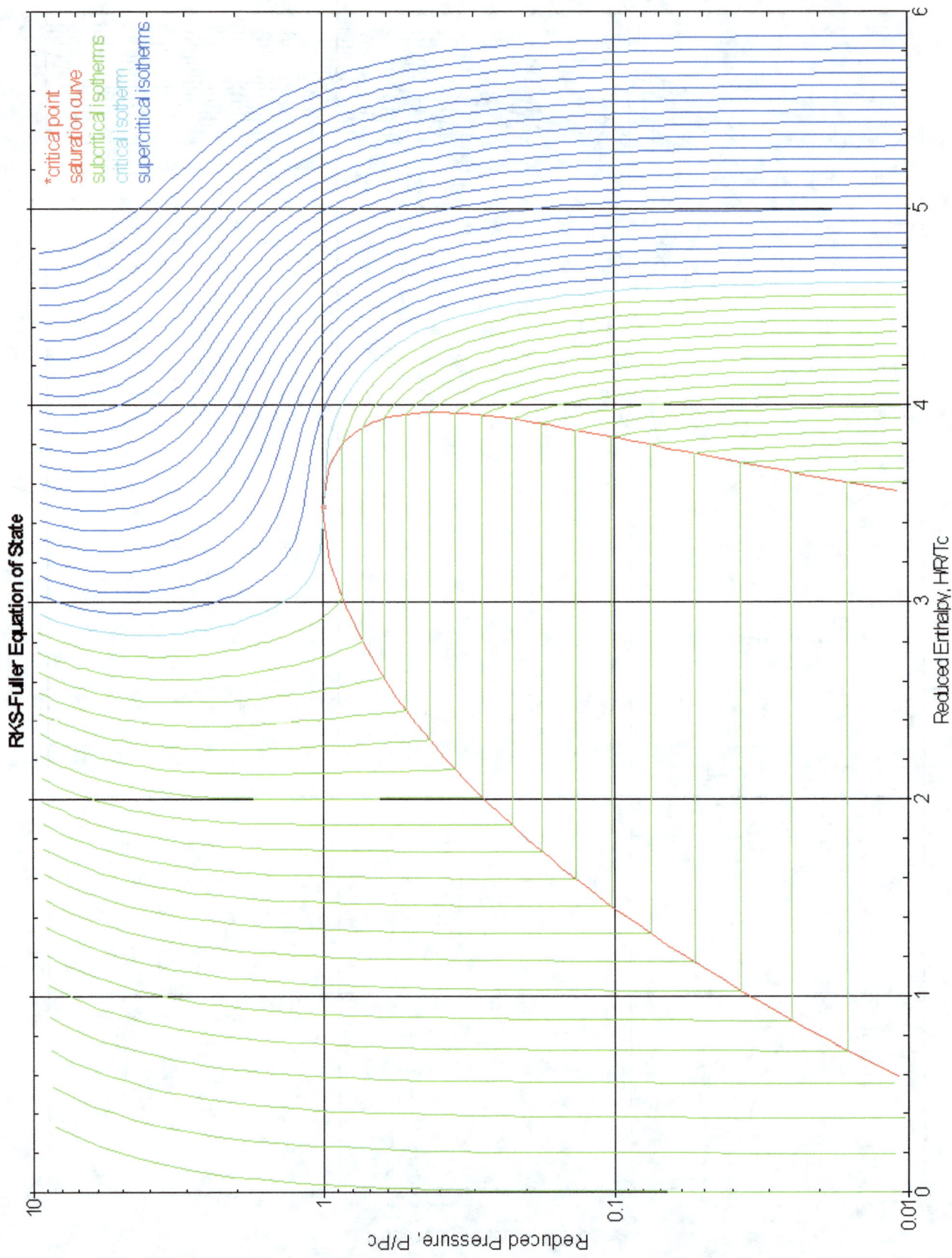

Figure 110. Pressure vs. Enthalpy Based on Fuller's Modification

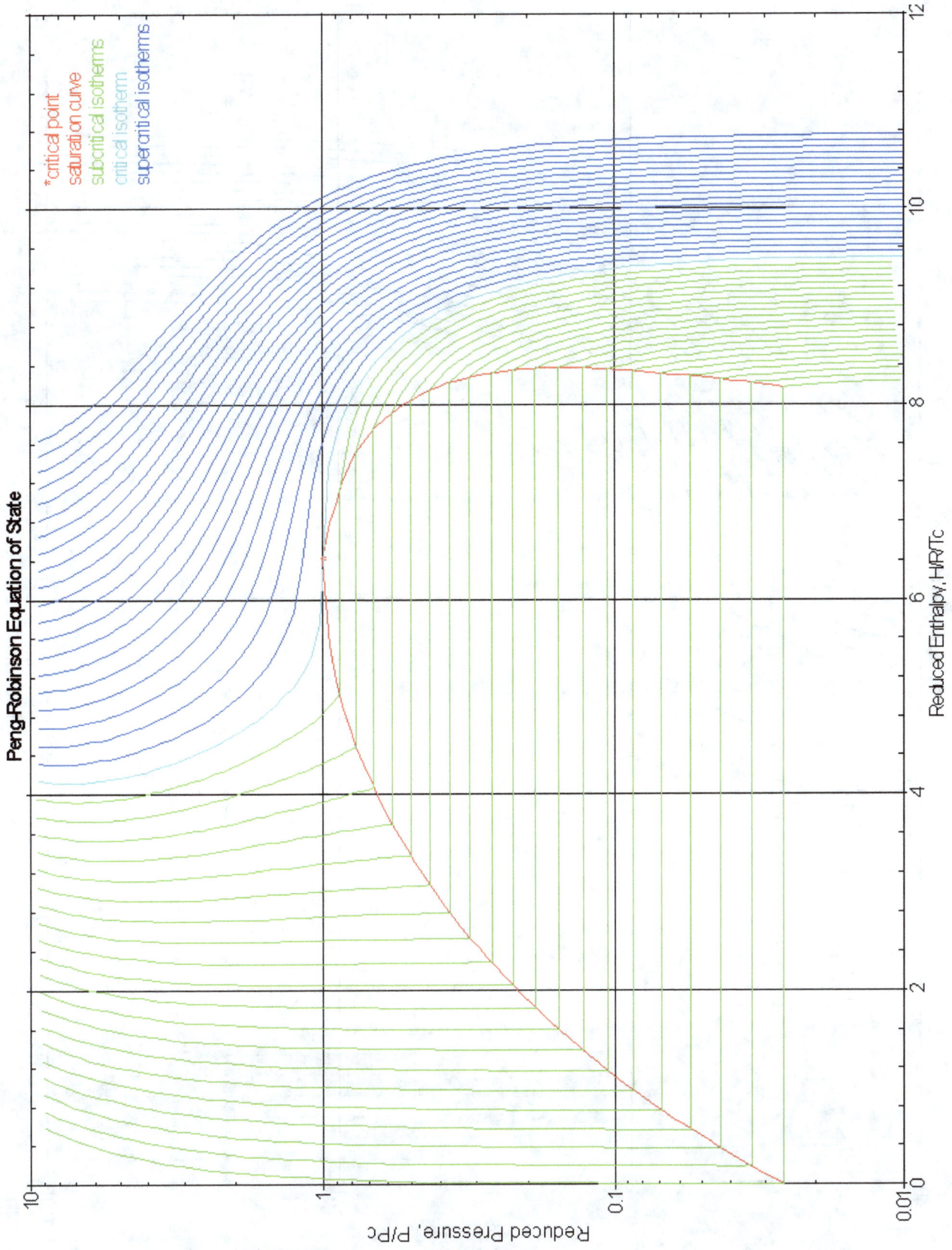

Figure 111. Pressure vs. Enthalpy Based on Peng-Robinson

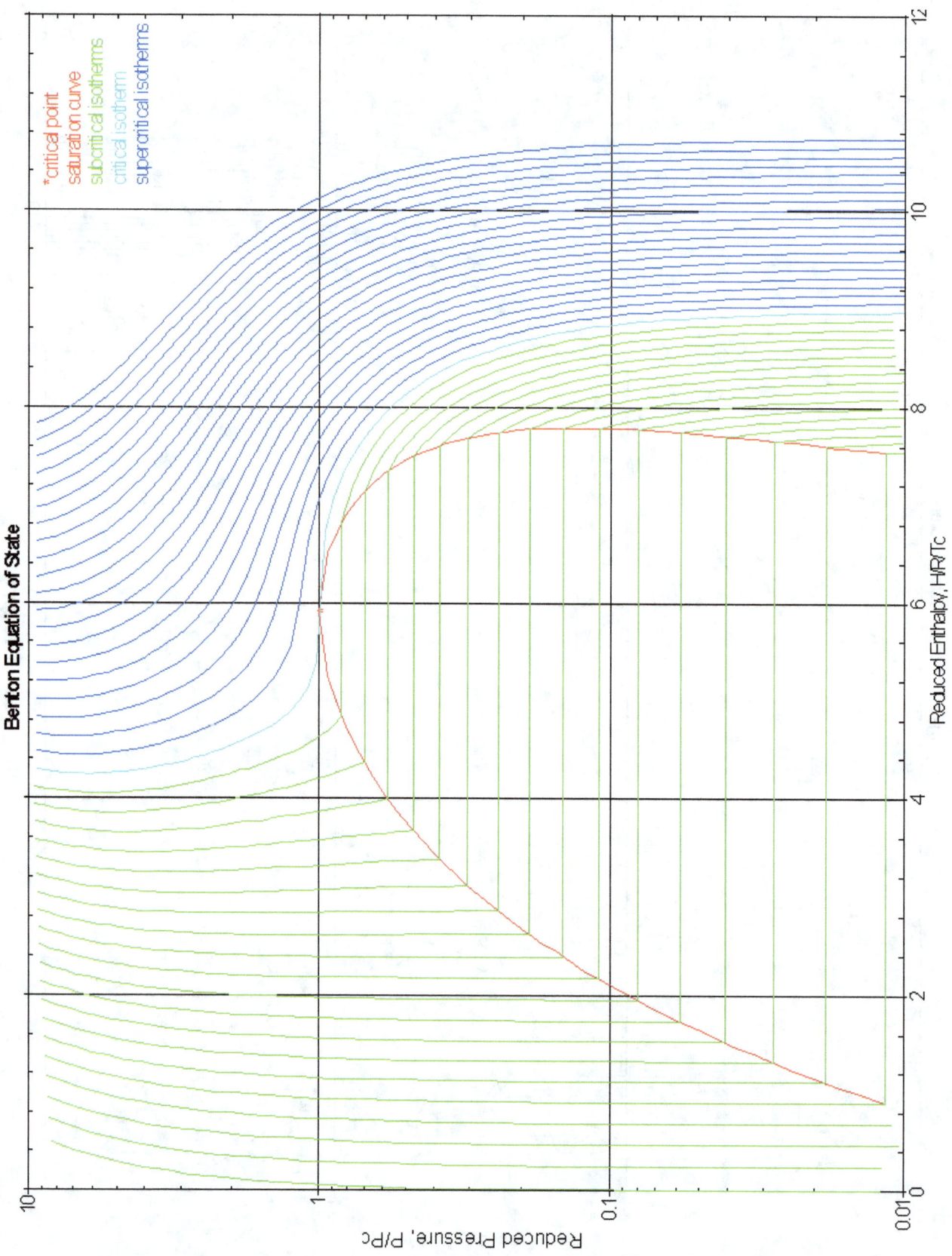

Figure 112. Pressure vs. Enthalpy Based on Author's Modification

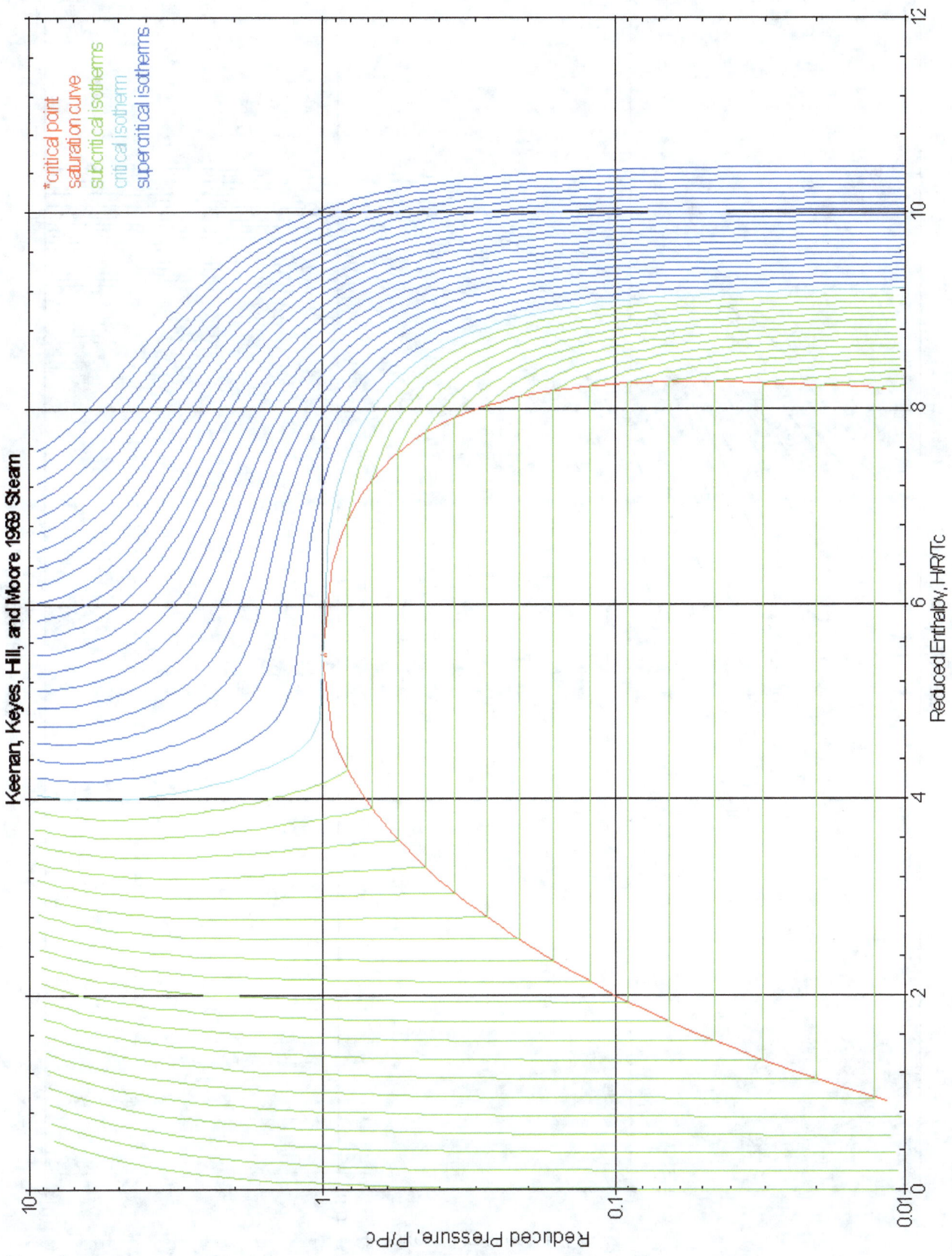

Figure 113. Pressure vs. Enthalpy Based on Keenan, Keyes, Hill, and Moore

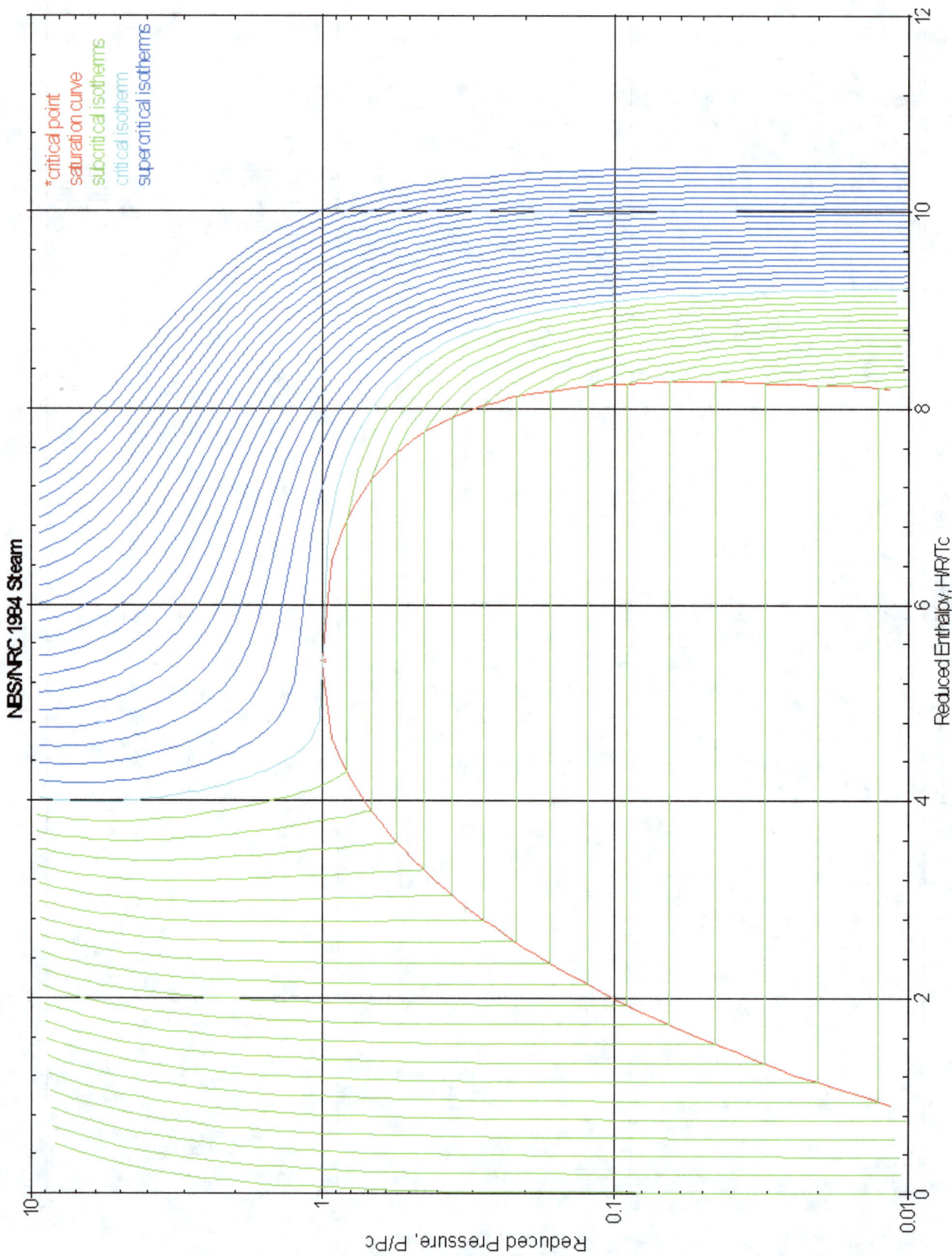

Figure 114. Pressure vs. Enthalpy Based on Haar, Gallagher, and Kell

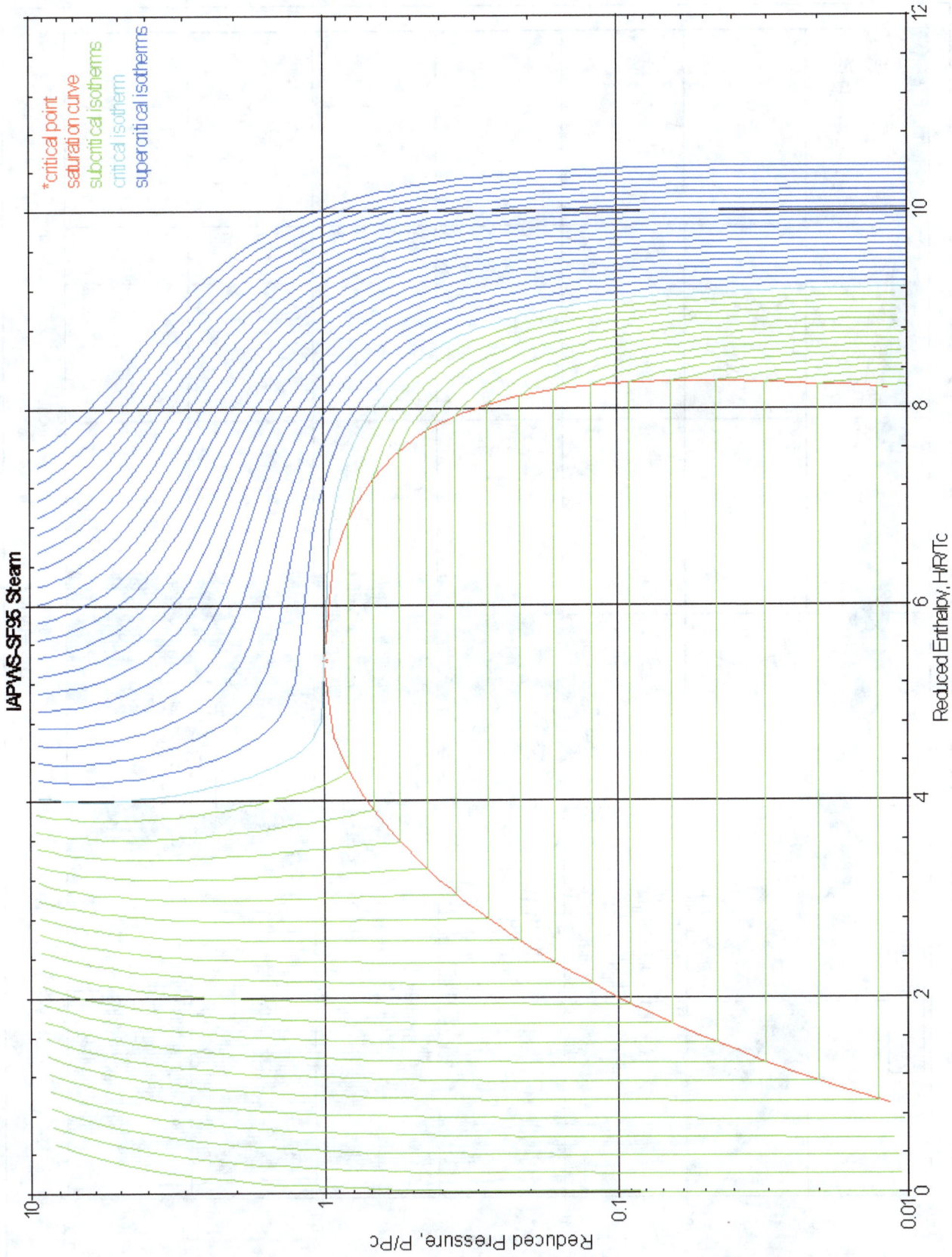

Figure 115. Pressure vs. Enthalpy Based on Wagner and Pruß

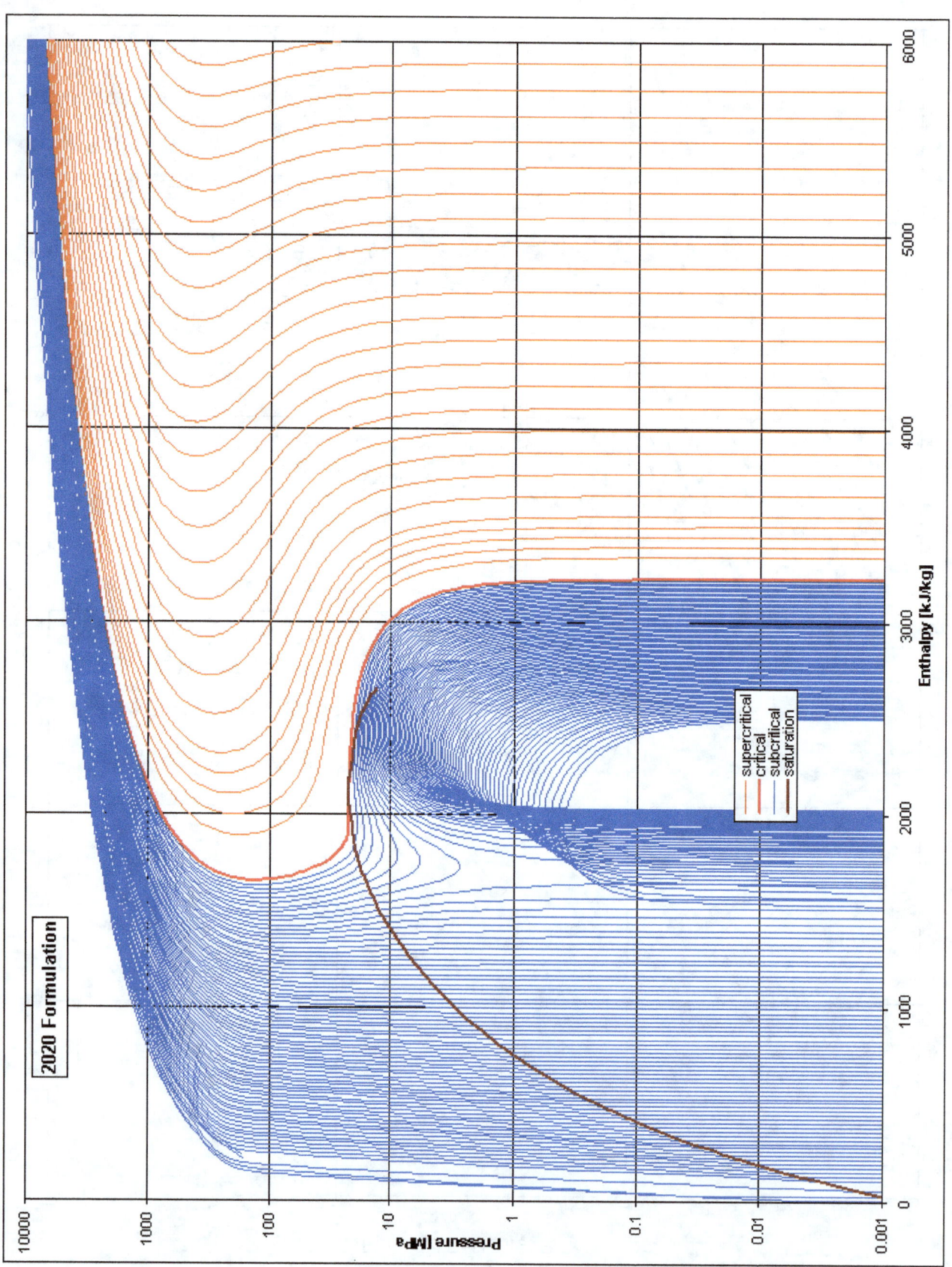

Figure 116. Pressure vs. Enthalpy Based on Steam 2020 Formulation

Note that this figure shows the metastable region, while the others in this chapter do not.

Chapter 9. Temperature vs. Entropy

A graph of T vs. S is often used in analyzing power cycles, as these are the same coordinates used to define the Carnot Cycle, which is has the highest possible thermal efficiency. We introduce no new equations in this chapter, only the graphs and so we begin with Nelson-Obert.

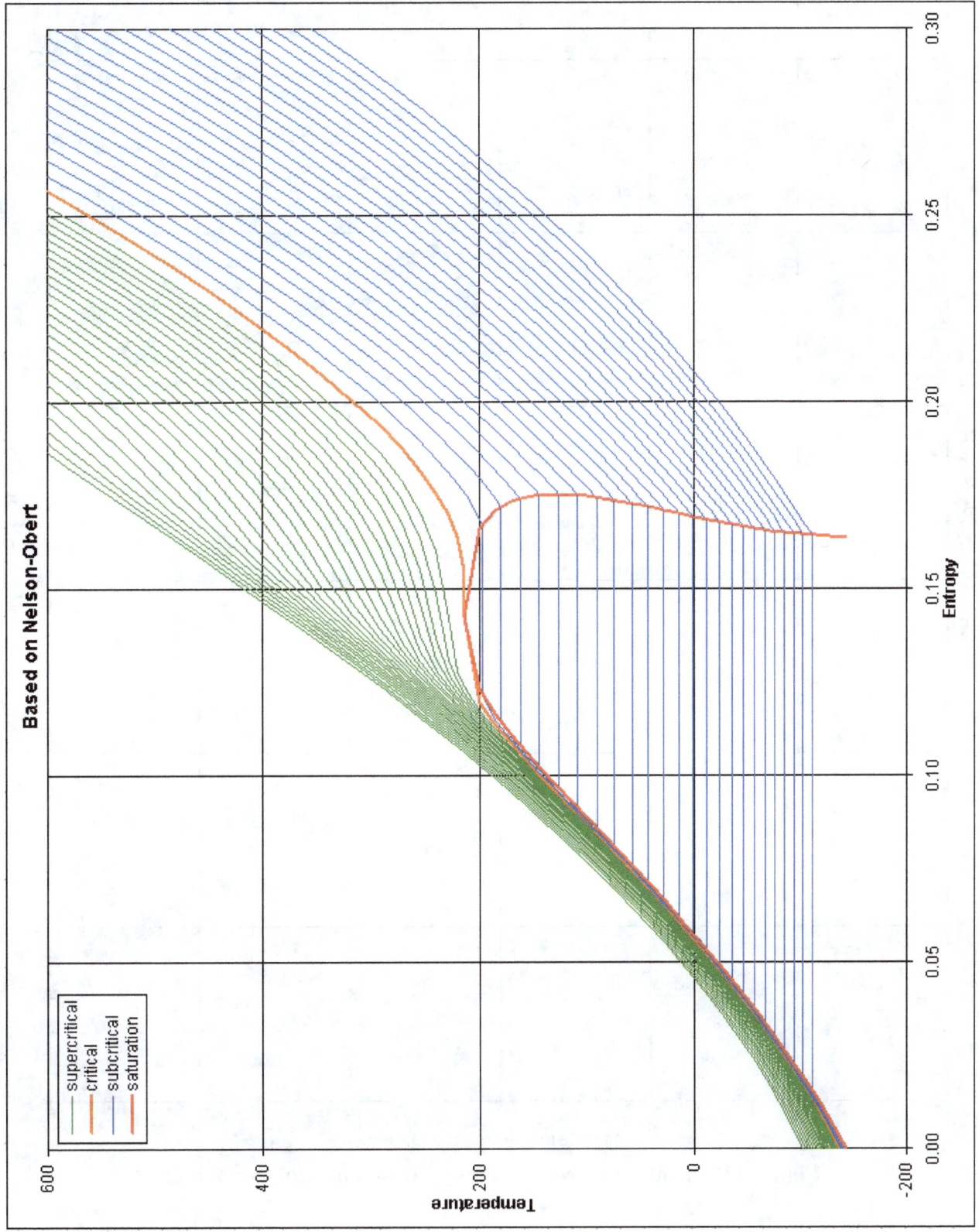

Figure 117. Temperature vs. Entropy Based on Nelson-Obert

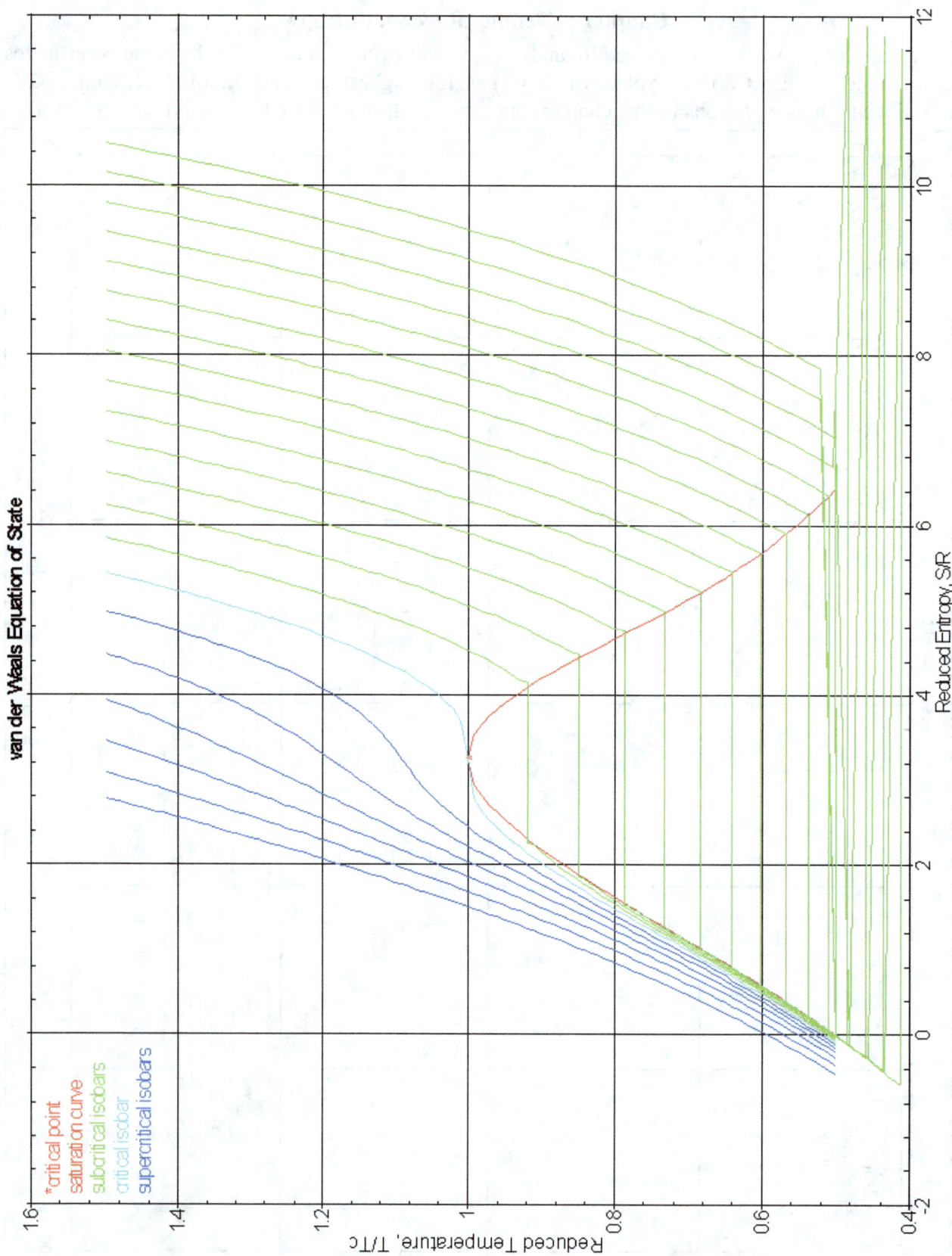

Figure 118. Temperature vs. Entropy Based on van der Waals

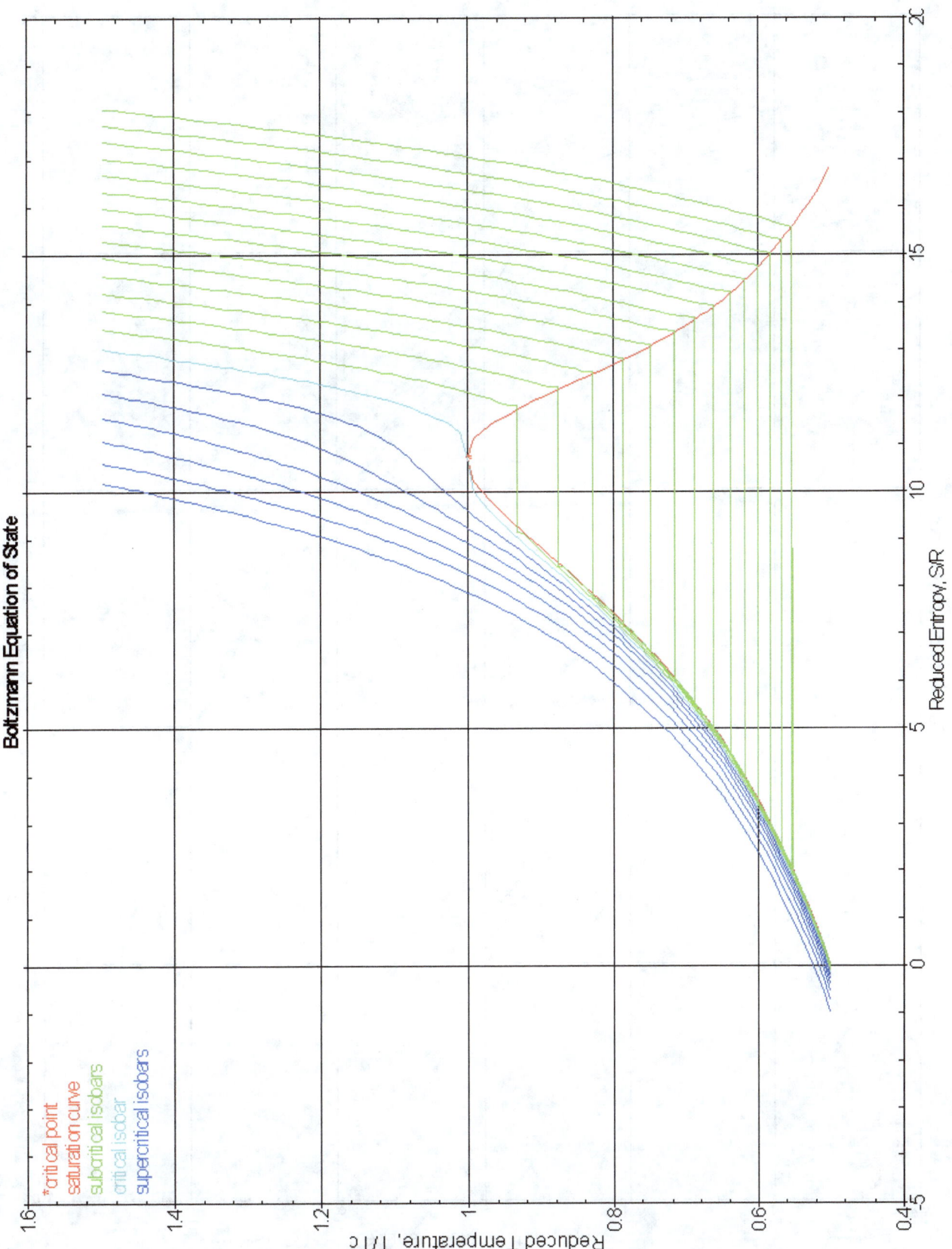

Figure 119. Temperature vs. Entropy Based on Boltzmann

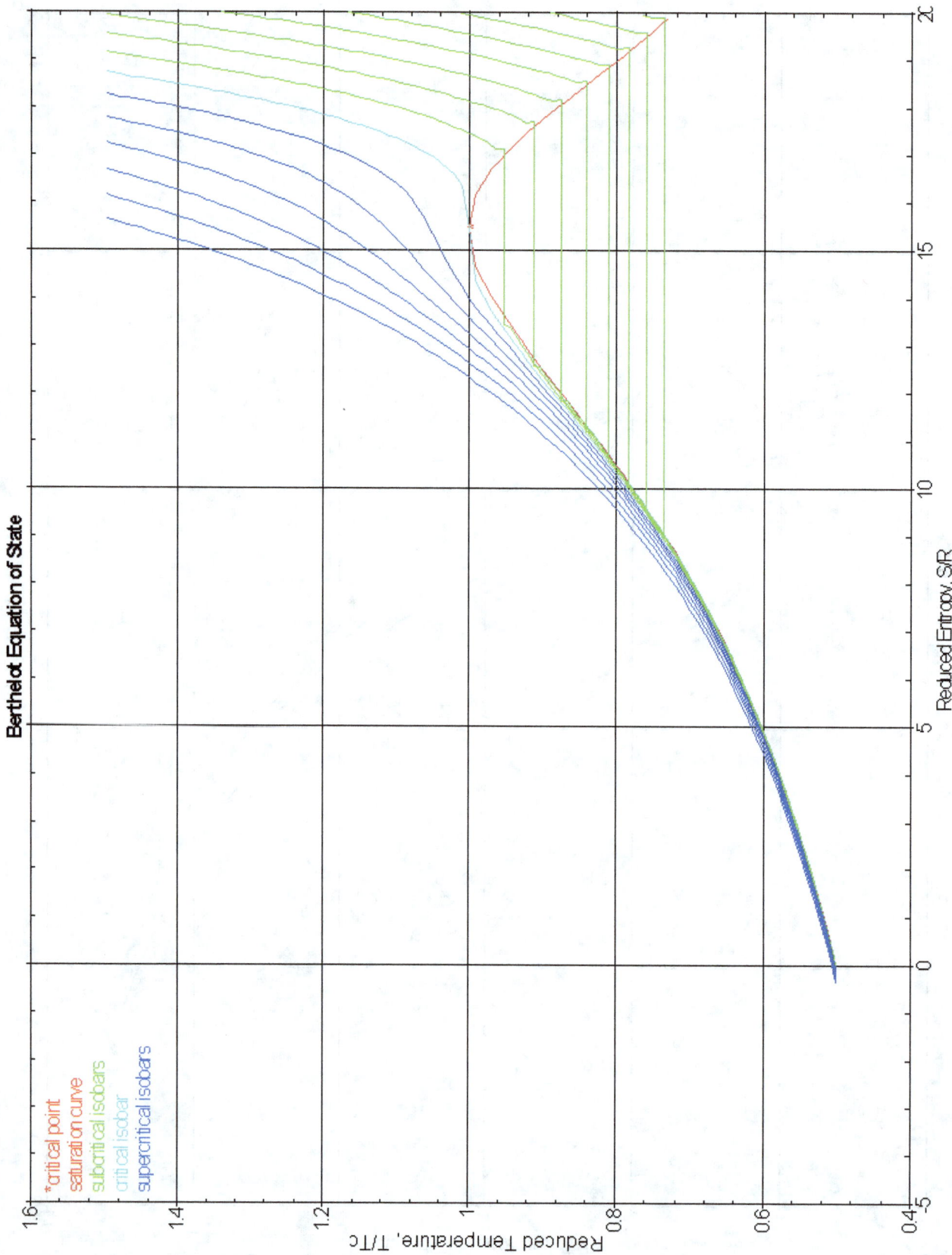

Figure 120. Temperature vs. Entropy Based on Berthelot

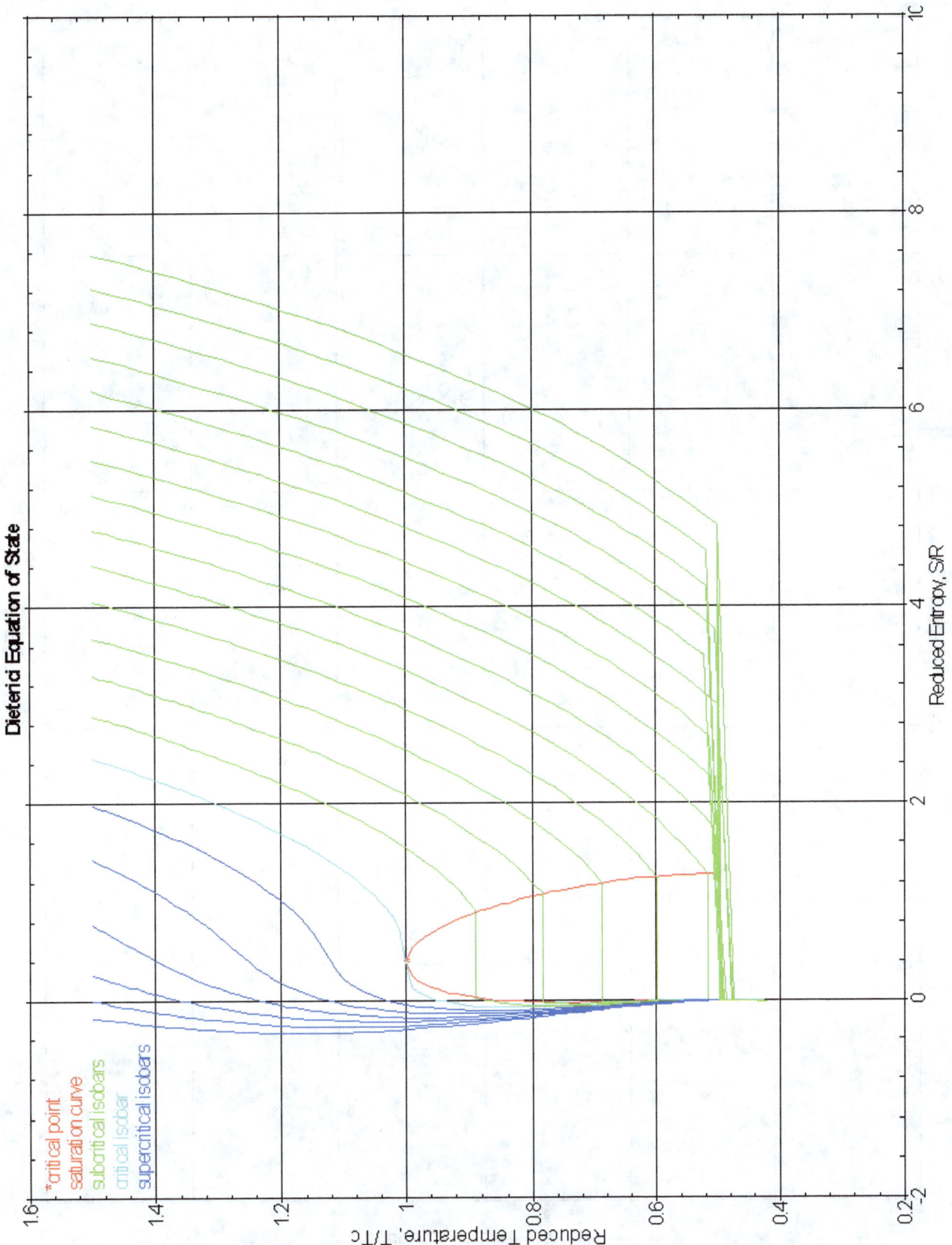

Figure 121. Temperature vs. Entropy Based on Dieterici

The behavior illustrated in the figure above is not acceptable.

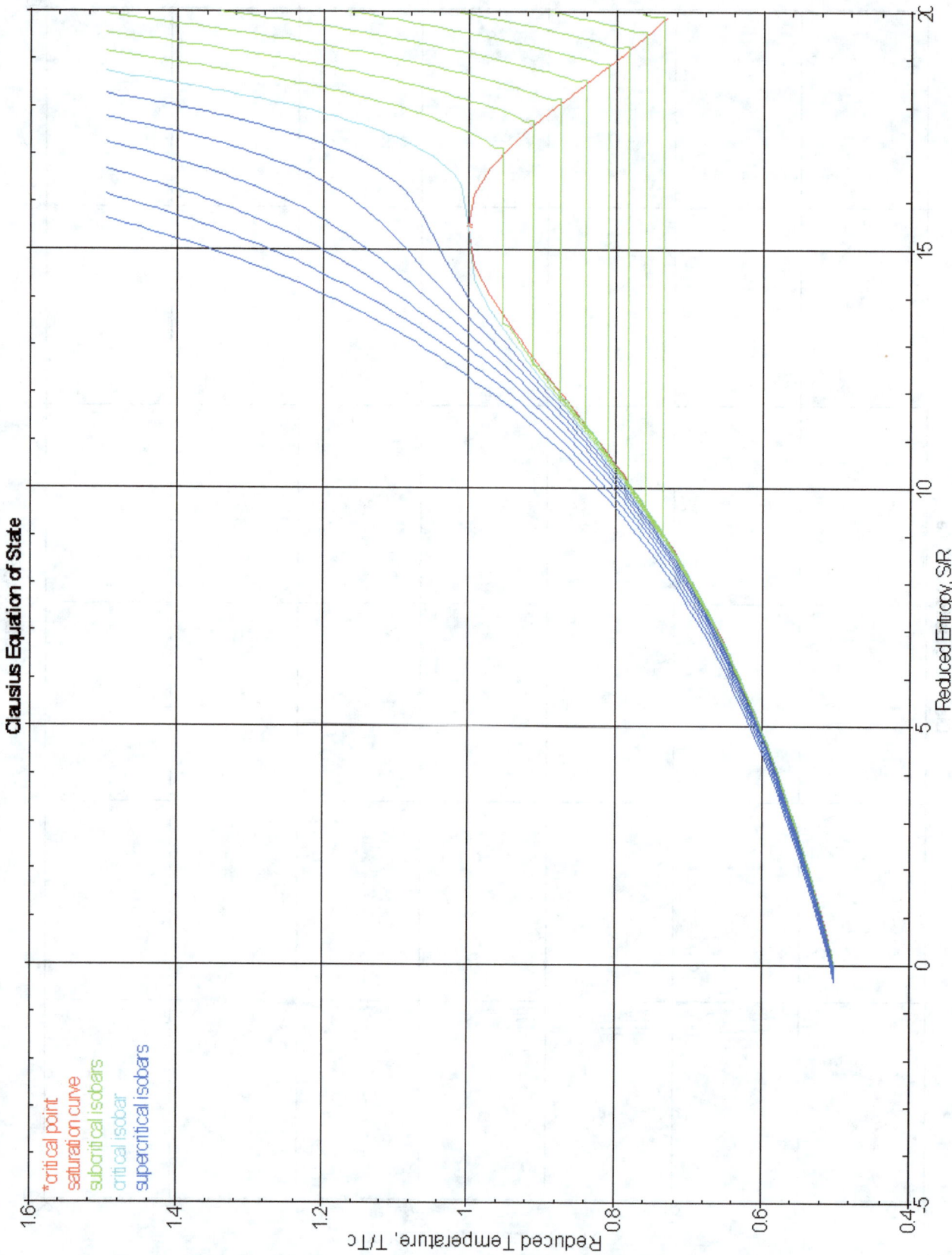

Figure 122. Temperature vs. Entropy Based on Clausius

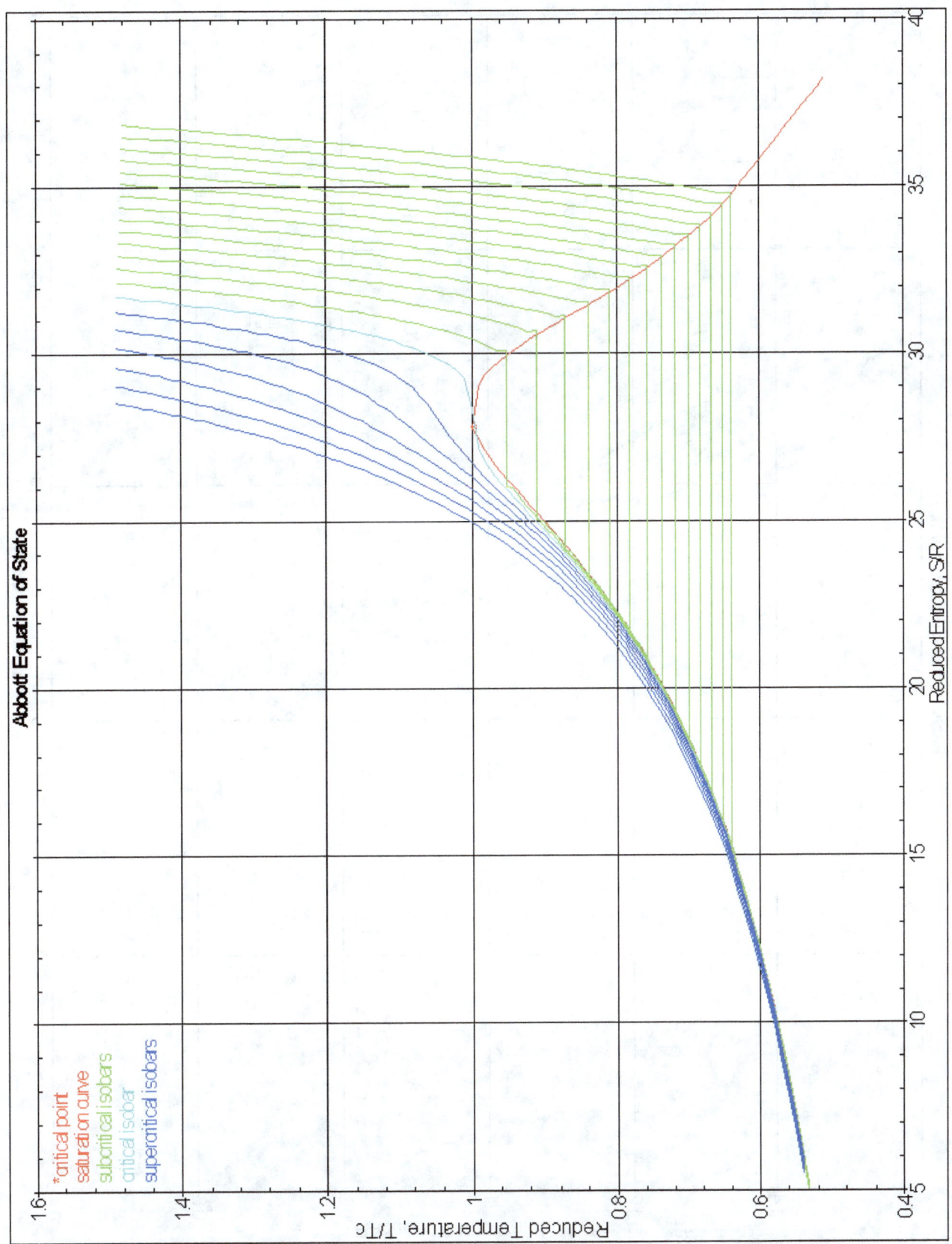

Figure 123. Temperature vs. Entropy Based on Abbott's Modification

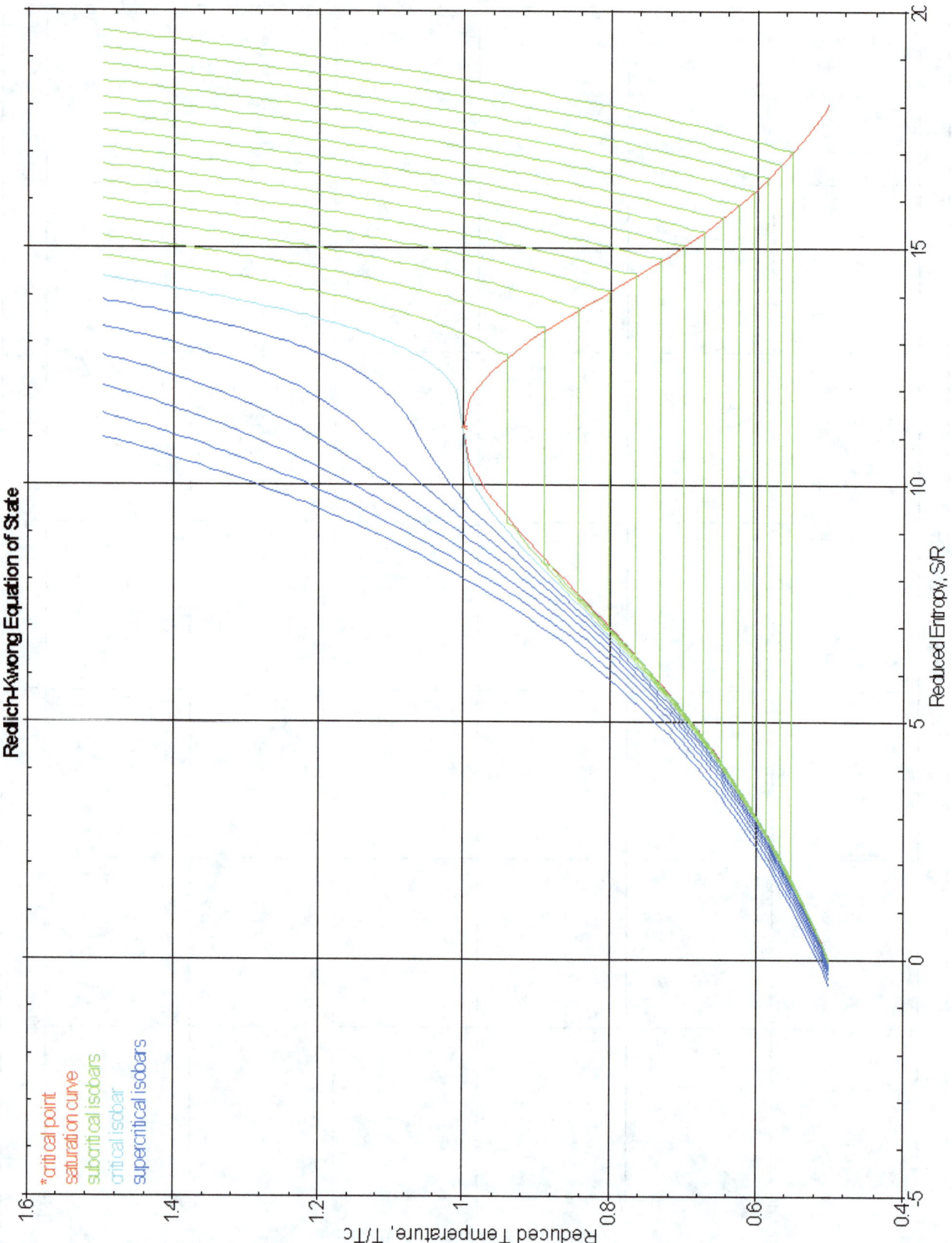

Figure 124. Temperature vs. Entropy Based on Redlich-Kwong

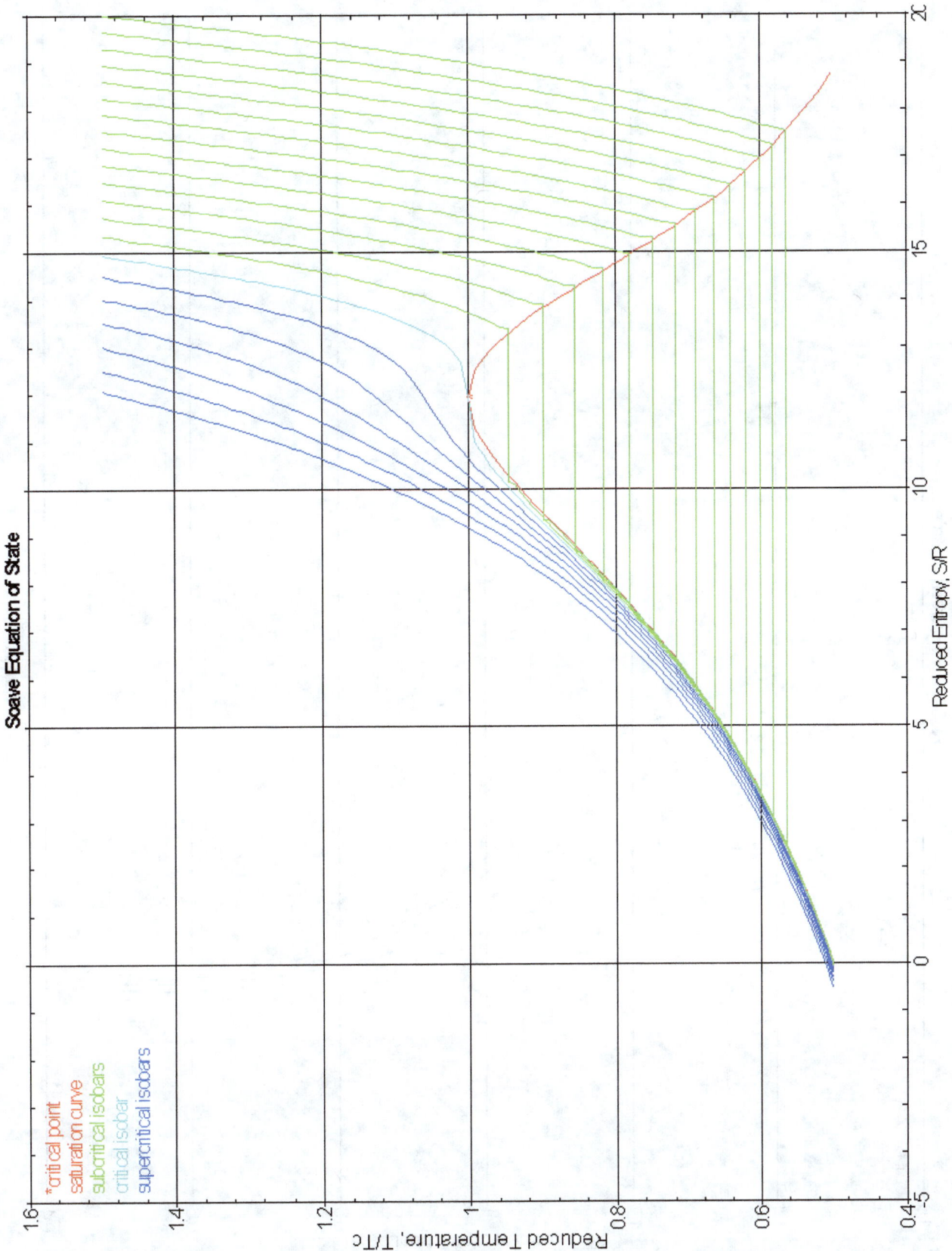

Figure 125. Temperature vs. Entropy Based on Soave's Modification

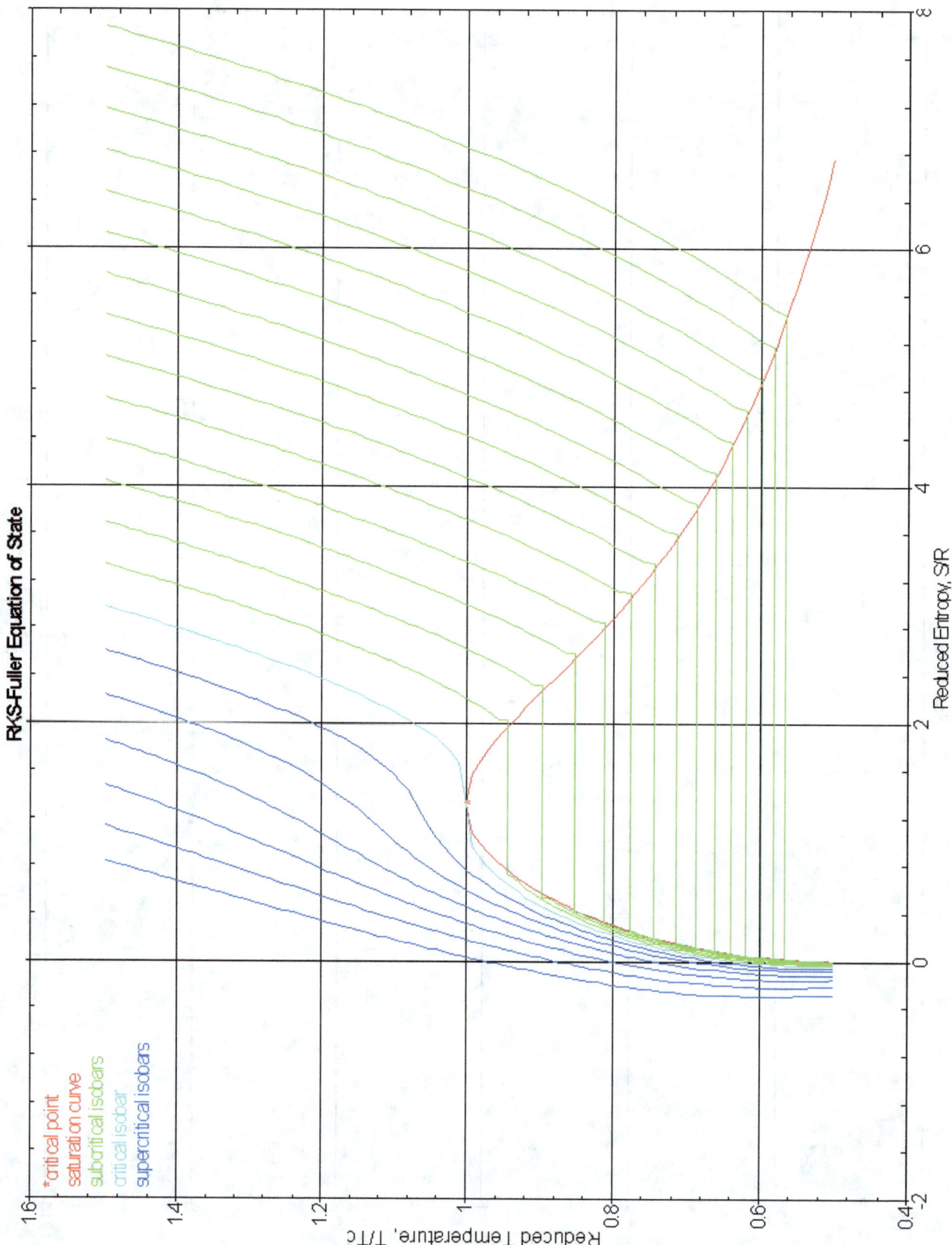

Figure 126. Temperature vs. Entropy Based on Fuller's Modification

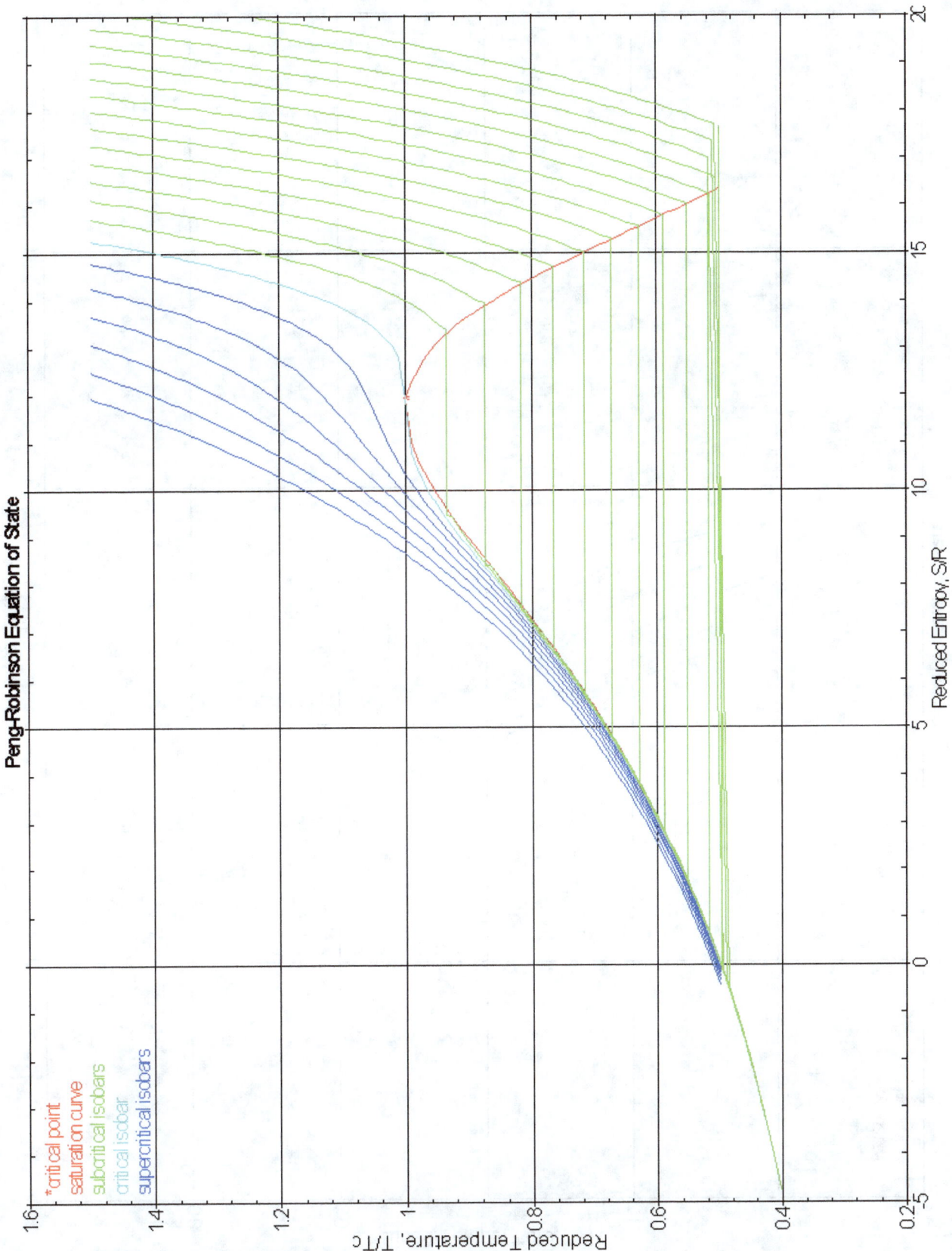

Figure 127. Temperature vs. Entropy Based on Peng-Robinson

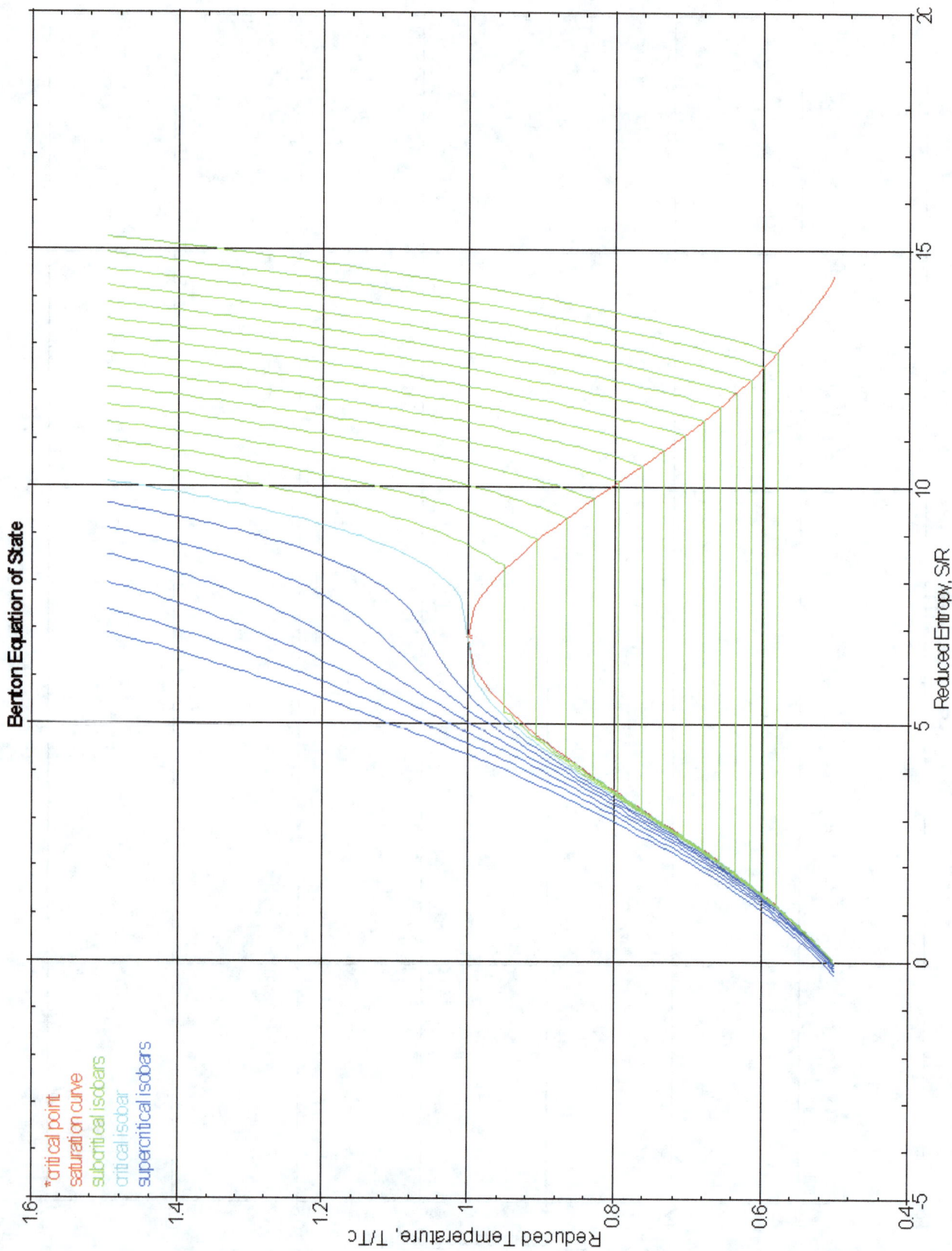

Figure 128. Temperature vs. Entropy Based on Author's Modification

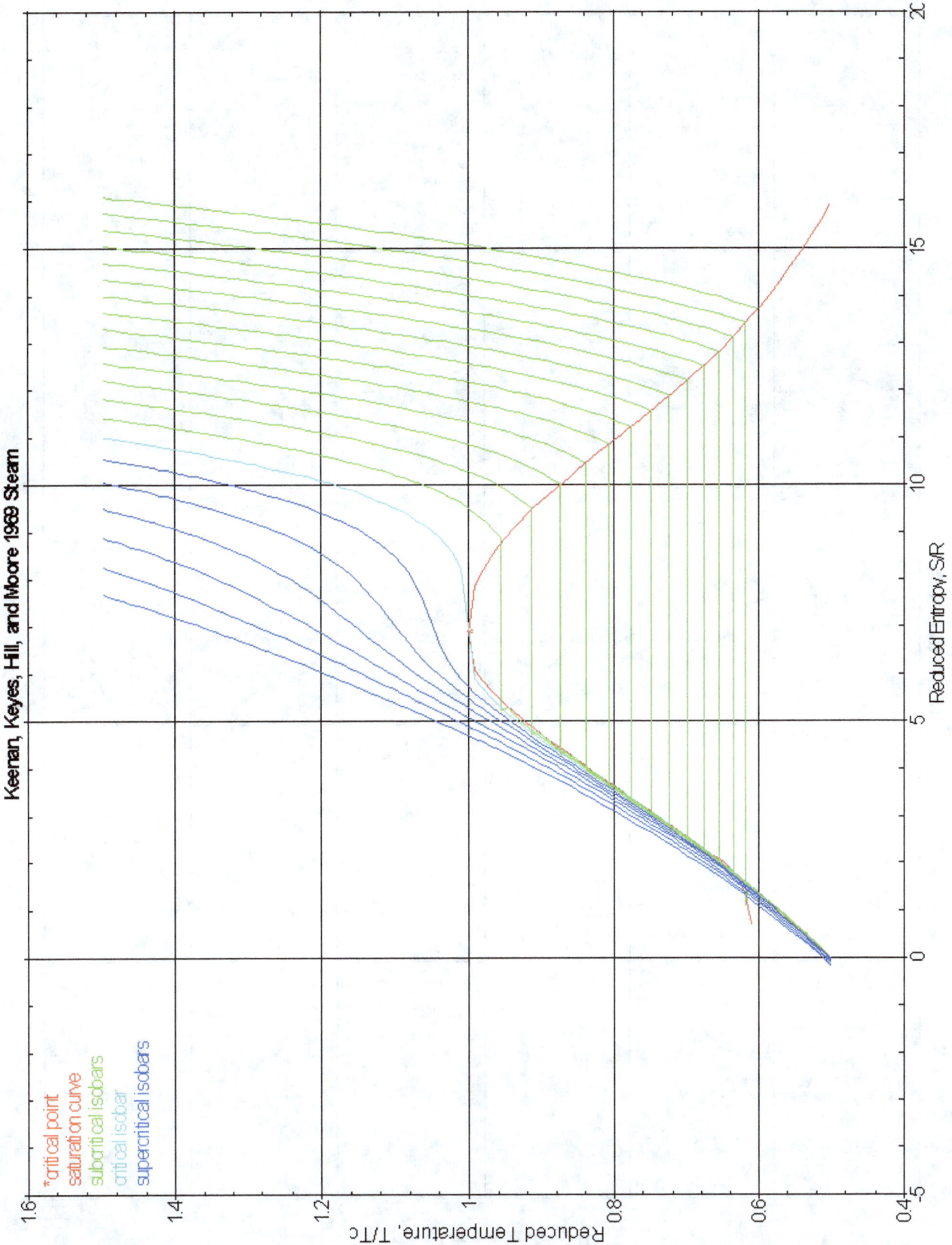

Figure 129. Temperature vs. Entropy Based on Keenan, Keyes, Hill, and Moore

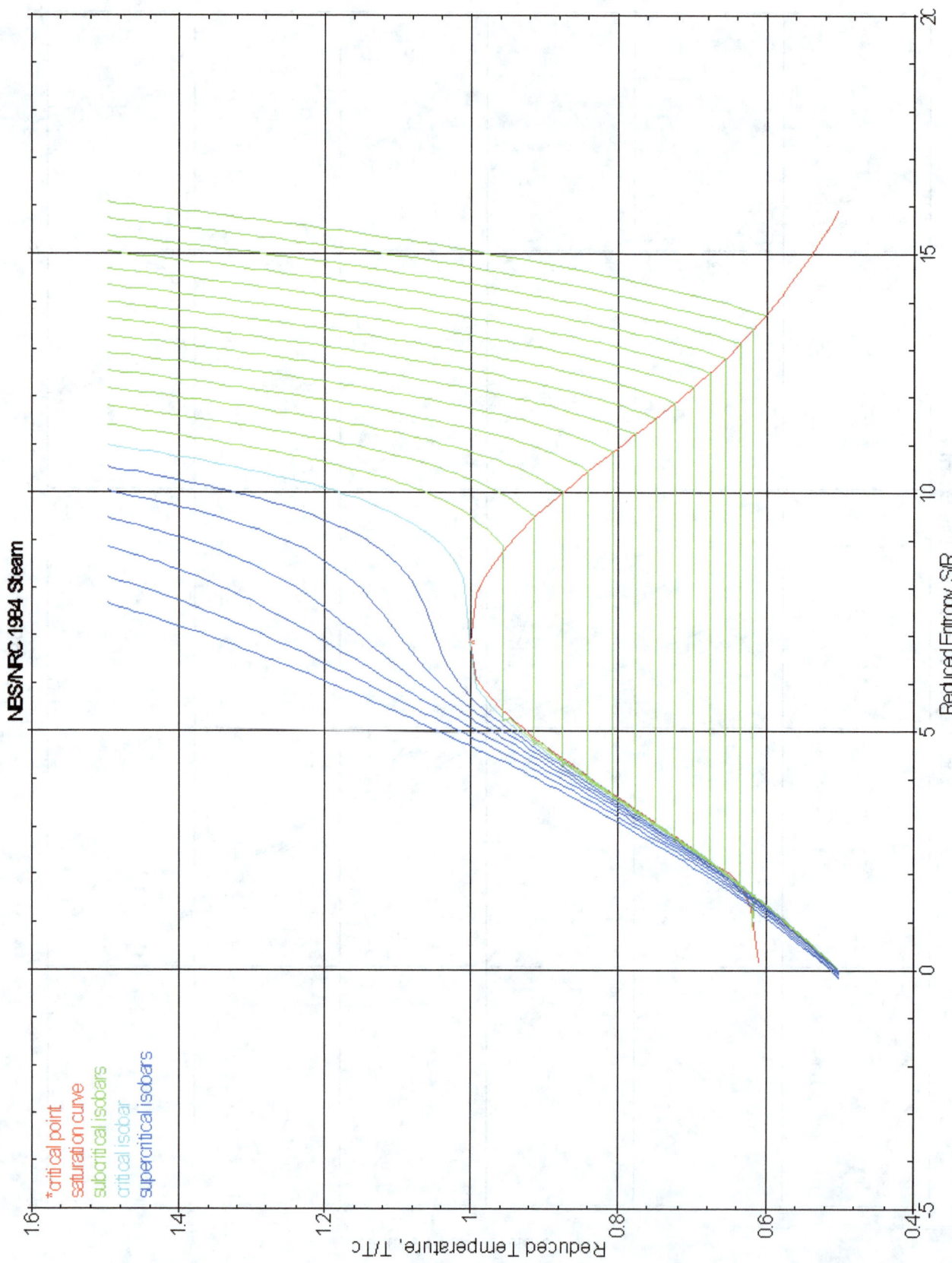

Figure 130. Temperature vs. Entropy Based on Haar, Gallagher, and Kell

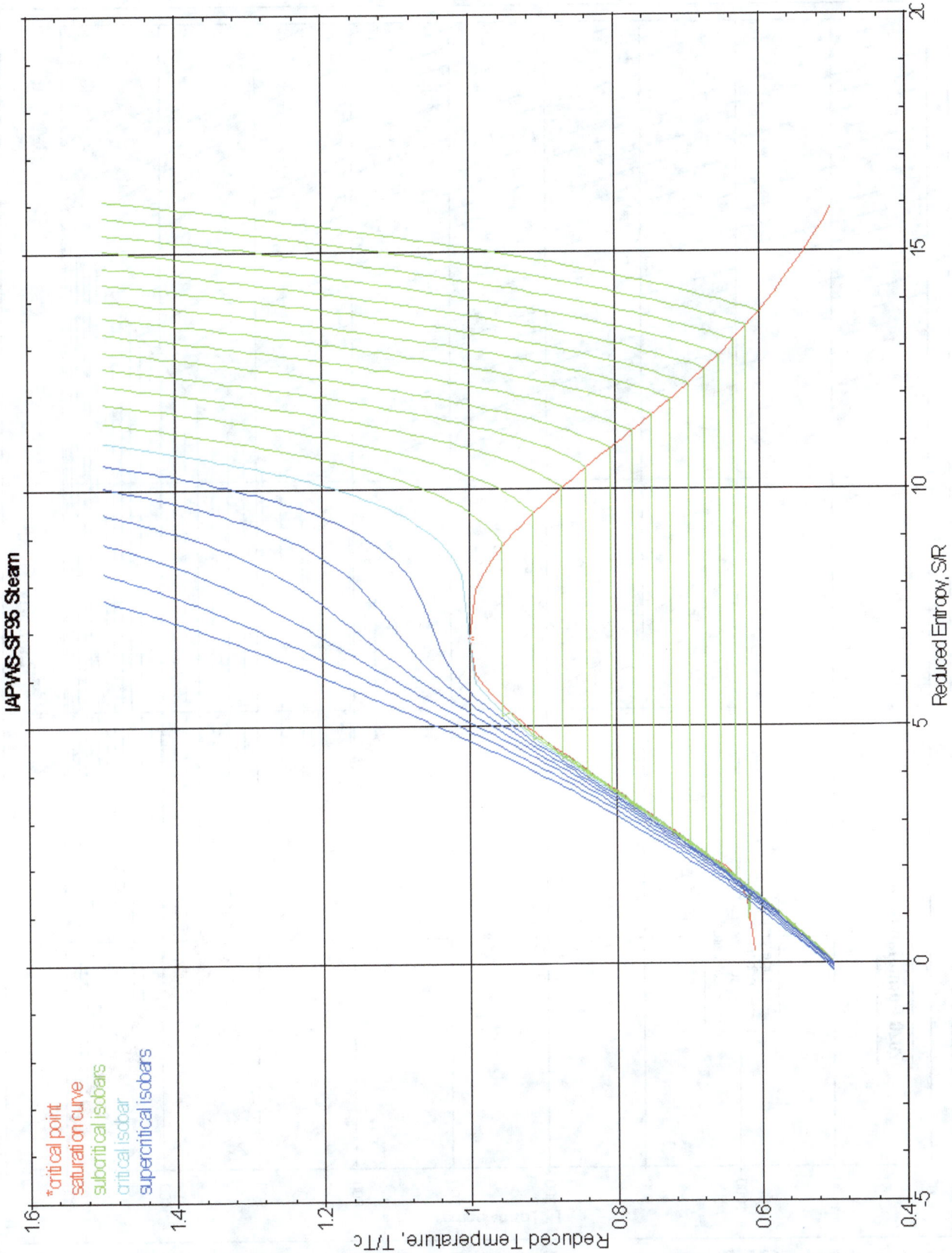

Figure 131. Temperature vs. Entropy Based on Wagner and Pruß

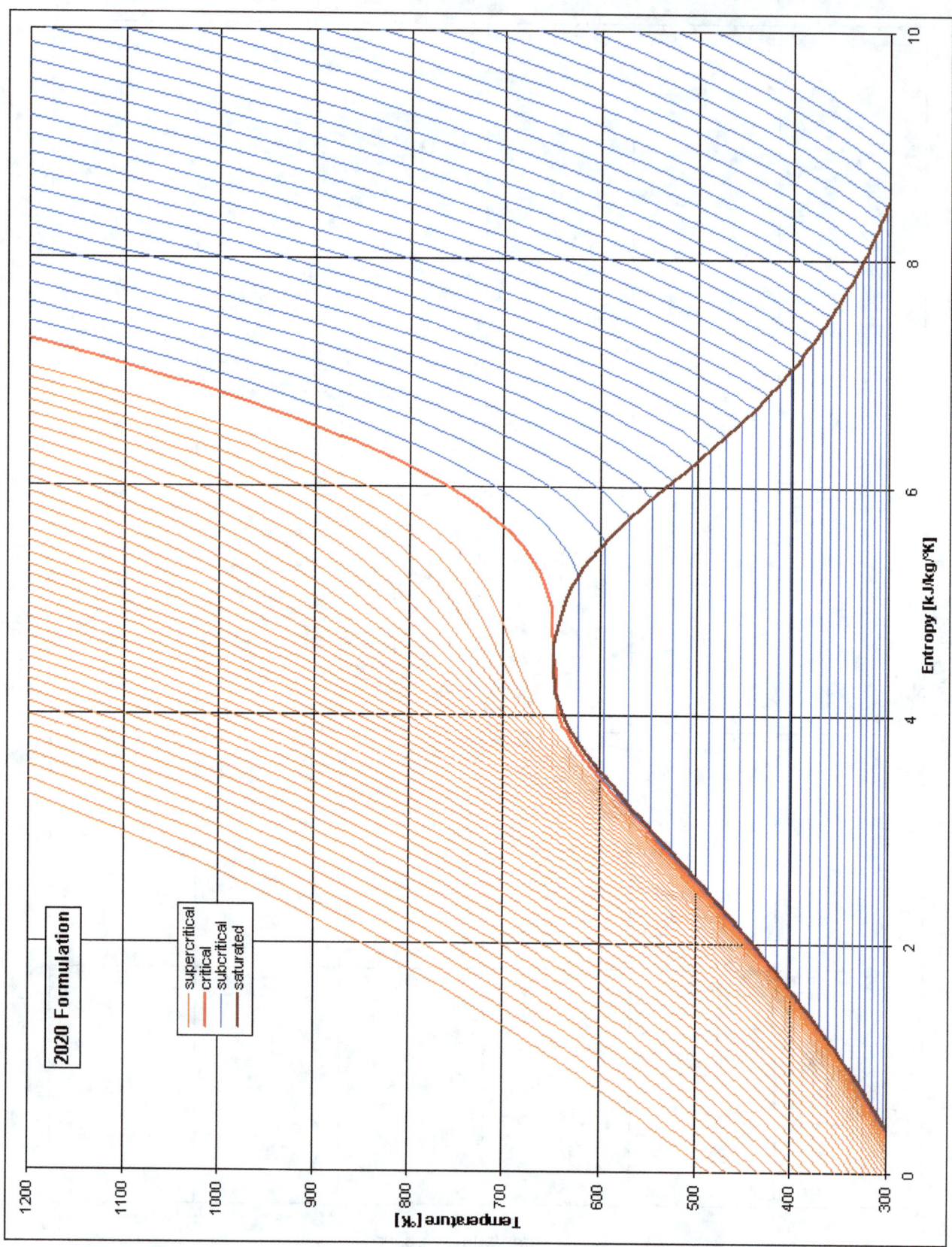

Figure 132. Temperature vs. Entropy Based on Steam 2020 Formulation

Chapter 10. Mollier Chart

A graph of H vs. S is called a Mollier Chart or Diagram. It is often used to illustrate processes in a power cycle, particularly expansion through a turbine. We do not introduce any new variables in this chapter and so will begin with the Nelson-Obert data:

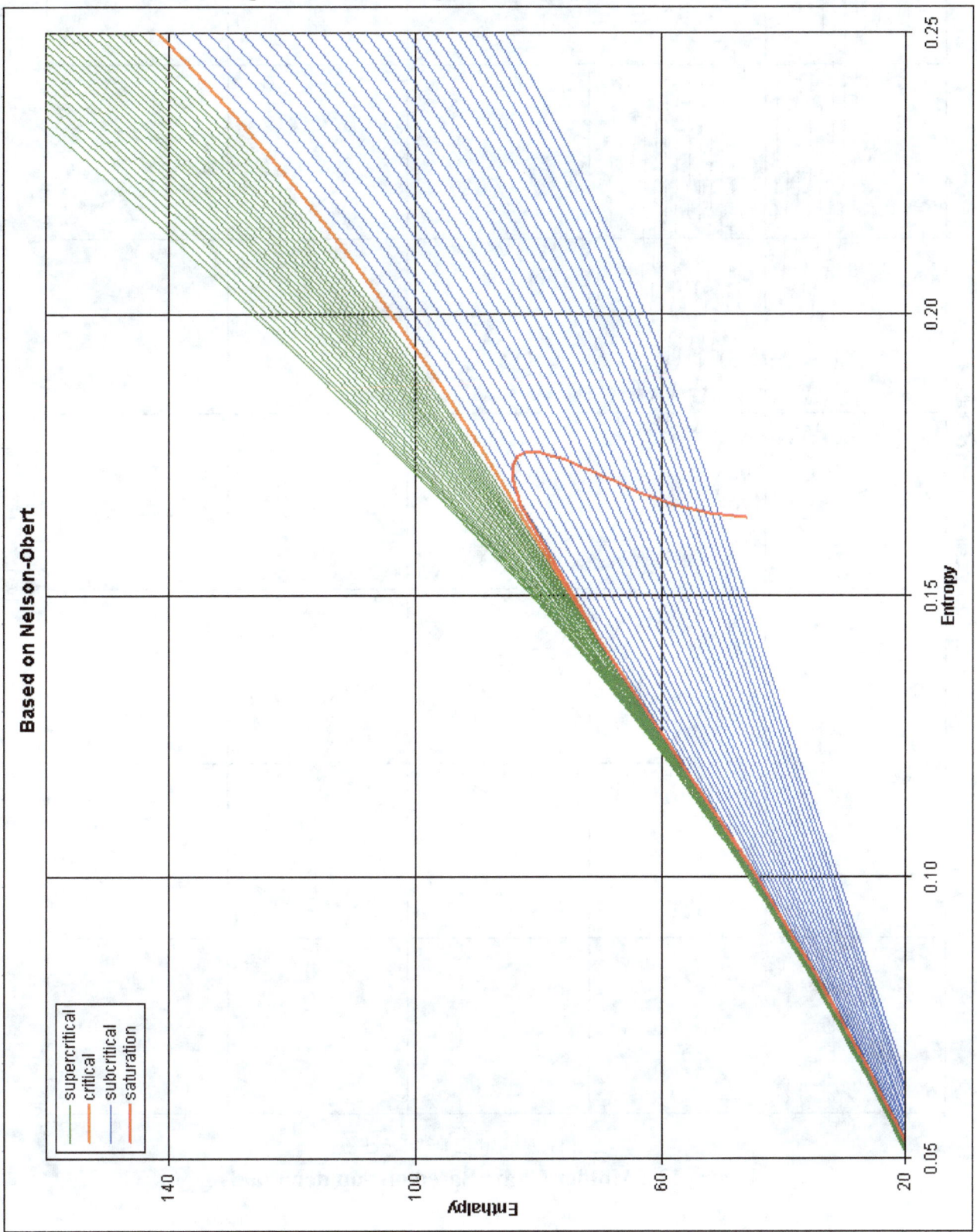

Figure 133. Mollier Chart Based on Nelson-Obert

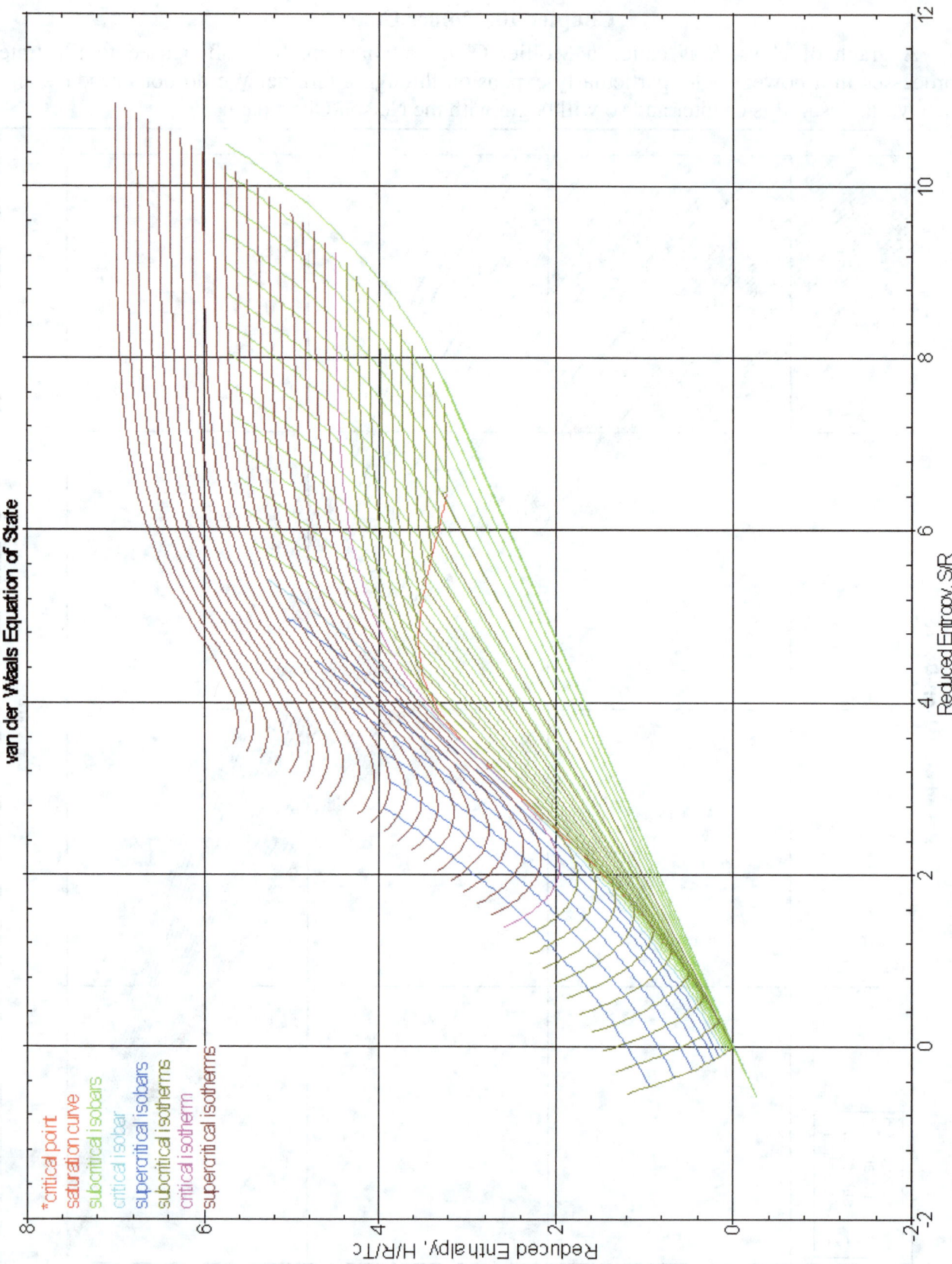

Figure 134. Mollier Chart Based on van der Waals

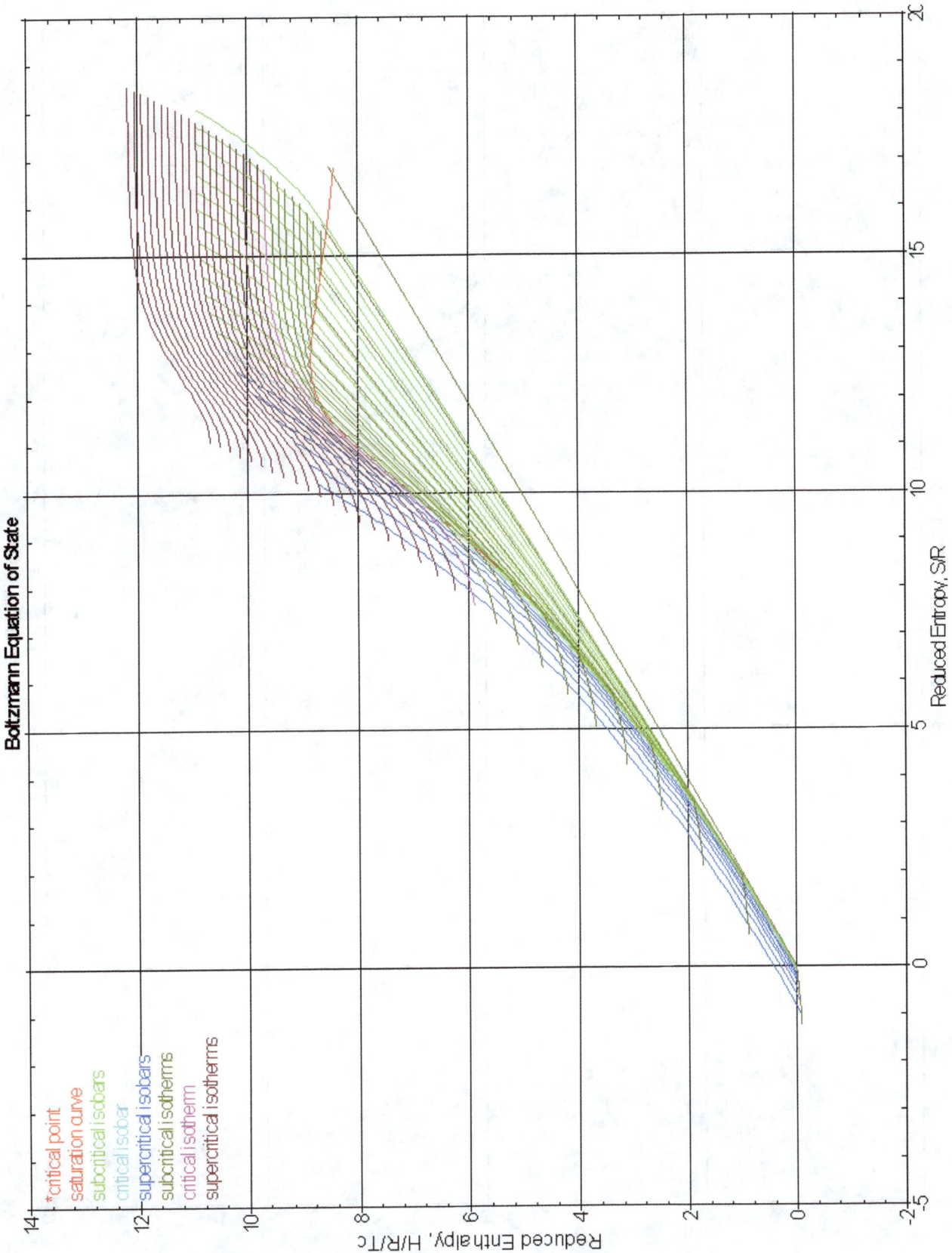

Figure 135. Mollier Chart Based on Boltzmann

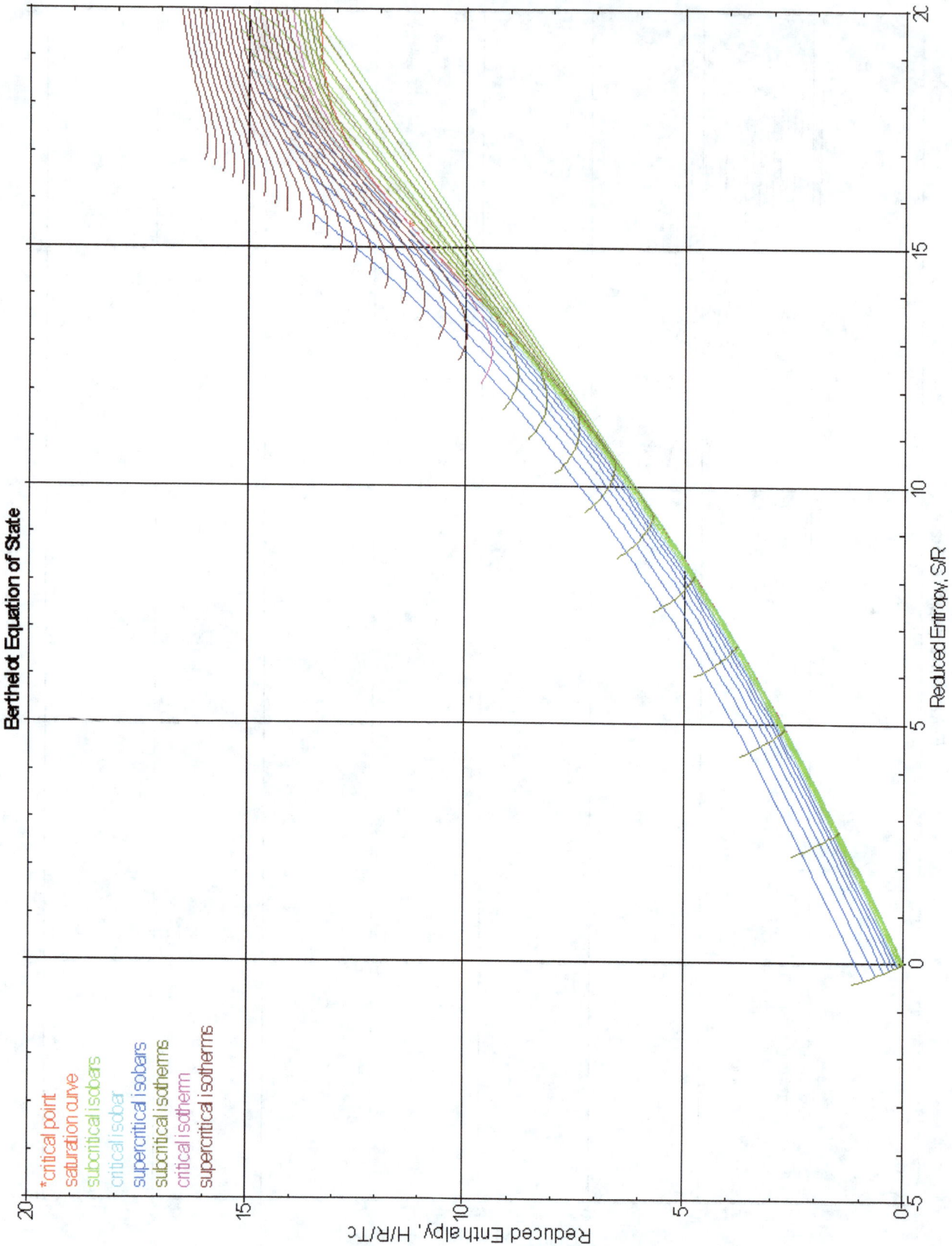

Figure 136. Mollier Chart Based on Berthelot

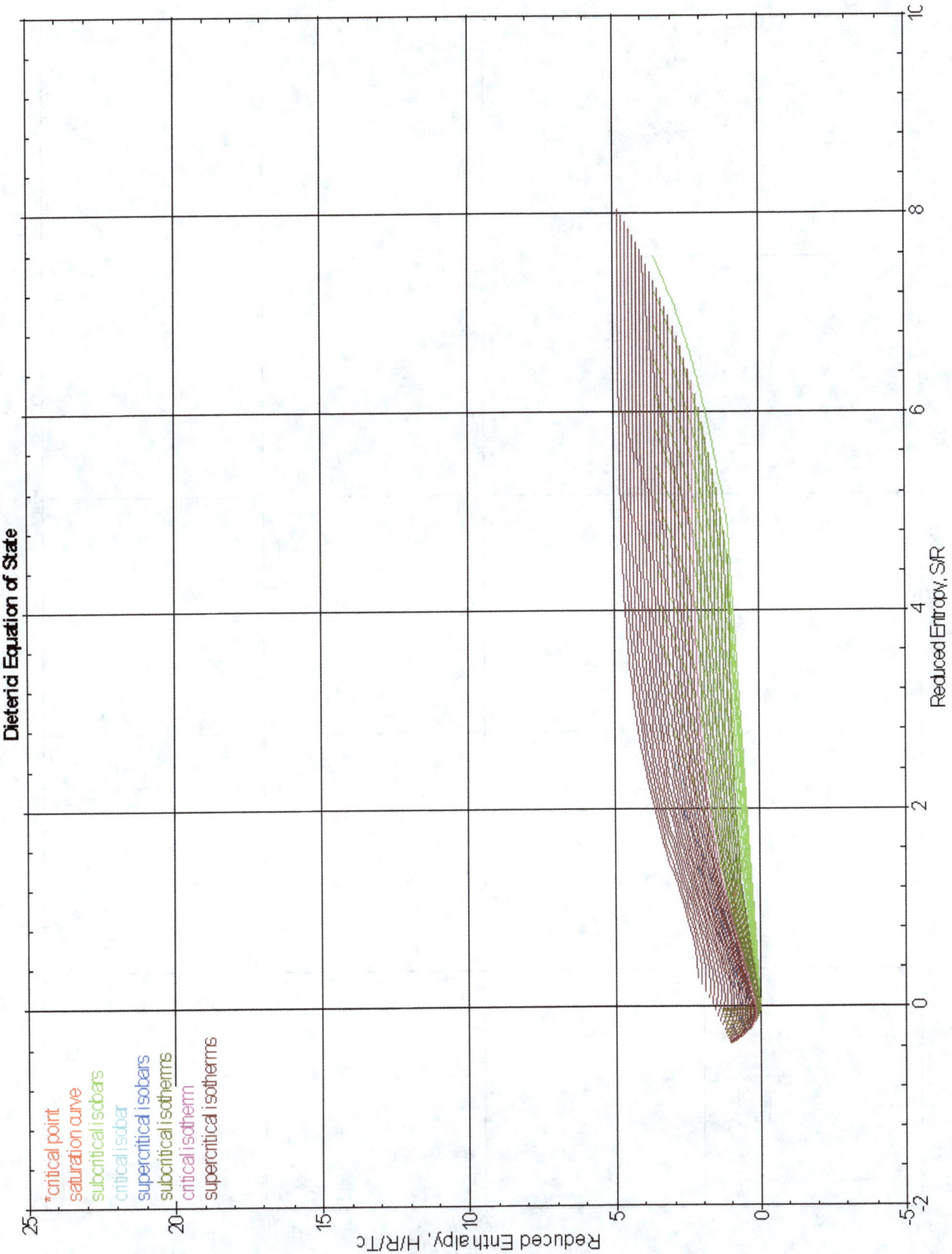

Figure 137. Mollier Chart Based on Dieterici

The behavior illustrated in the figure above is unacceptable.

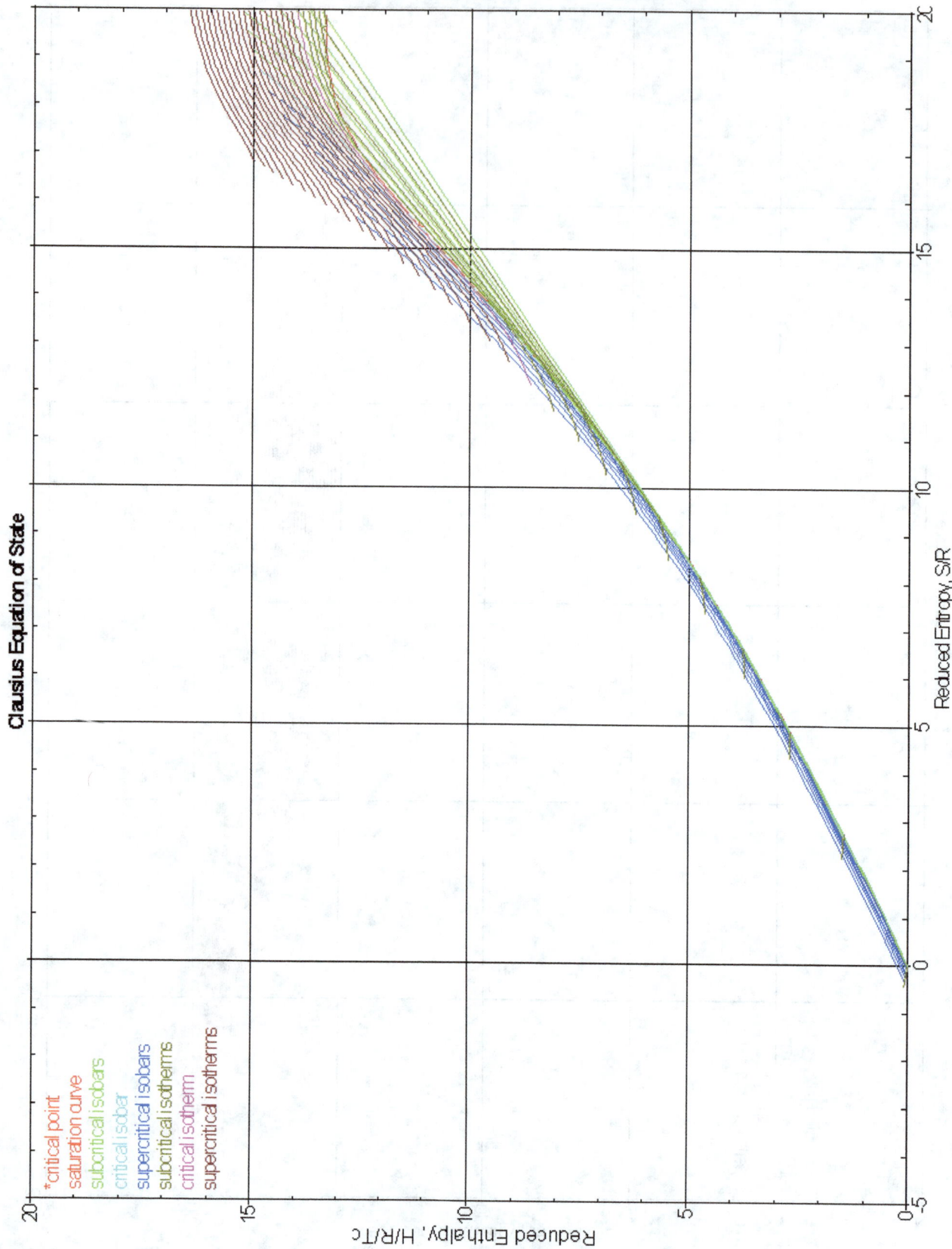

Figure 138. Mollier Chart Based on Clausius

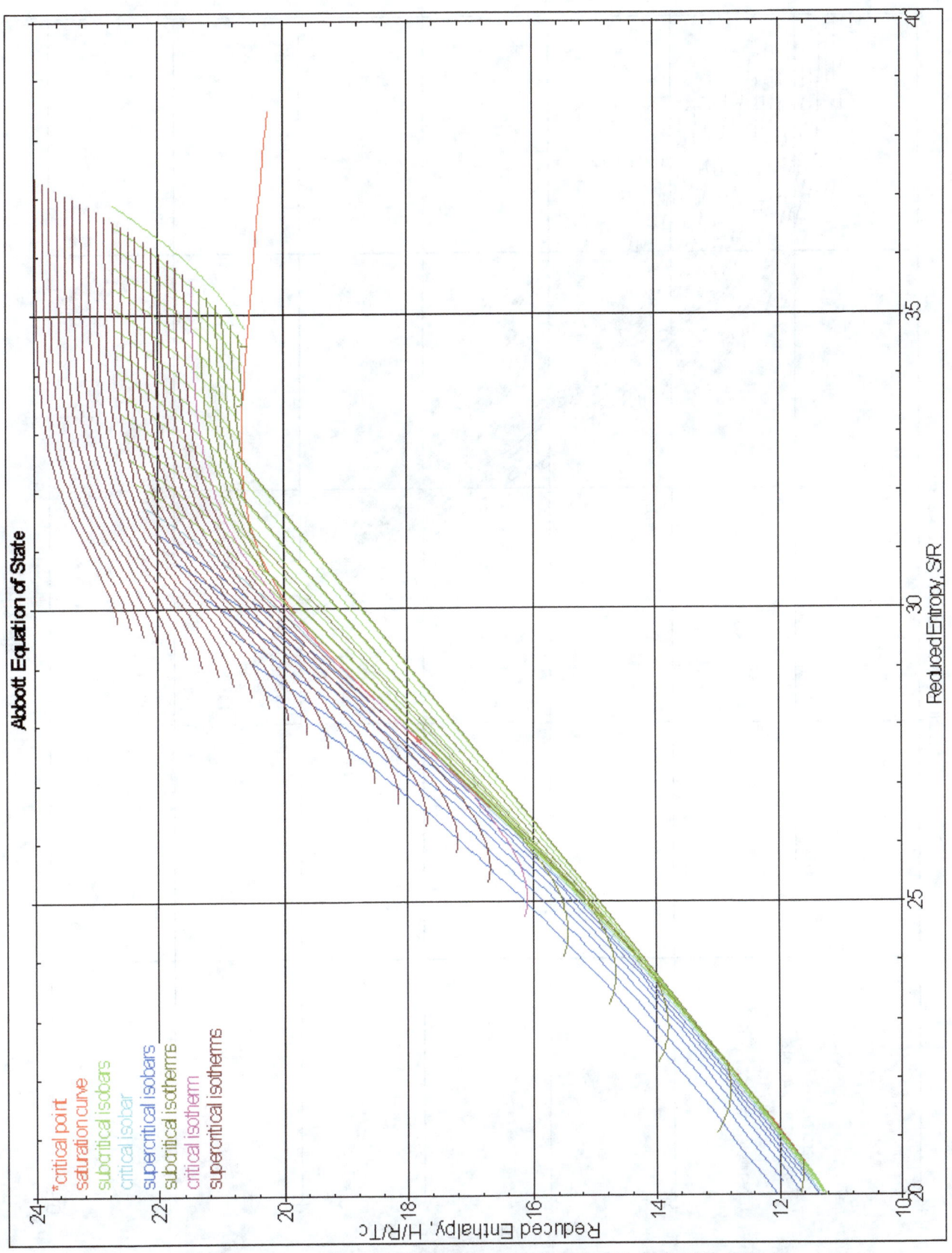

Figure 139. Mollier Chart Based on Abbott's Modification

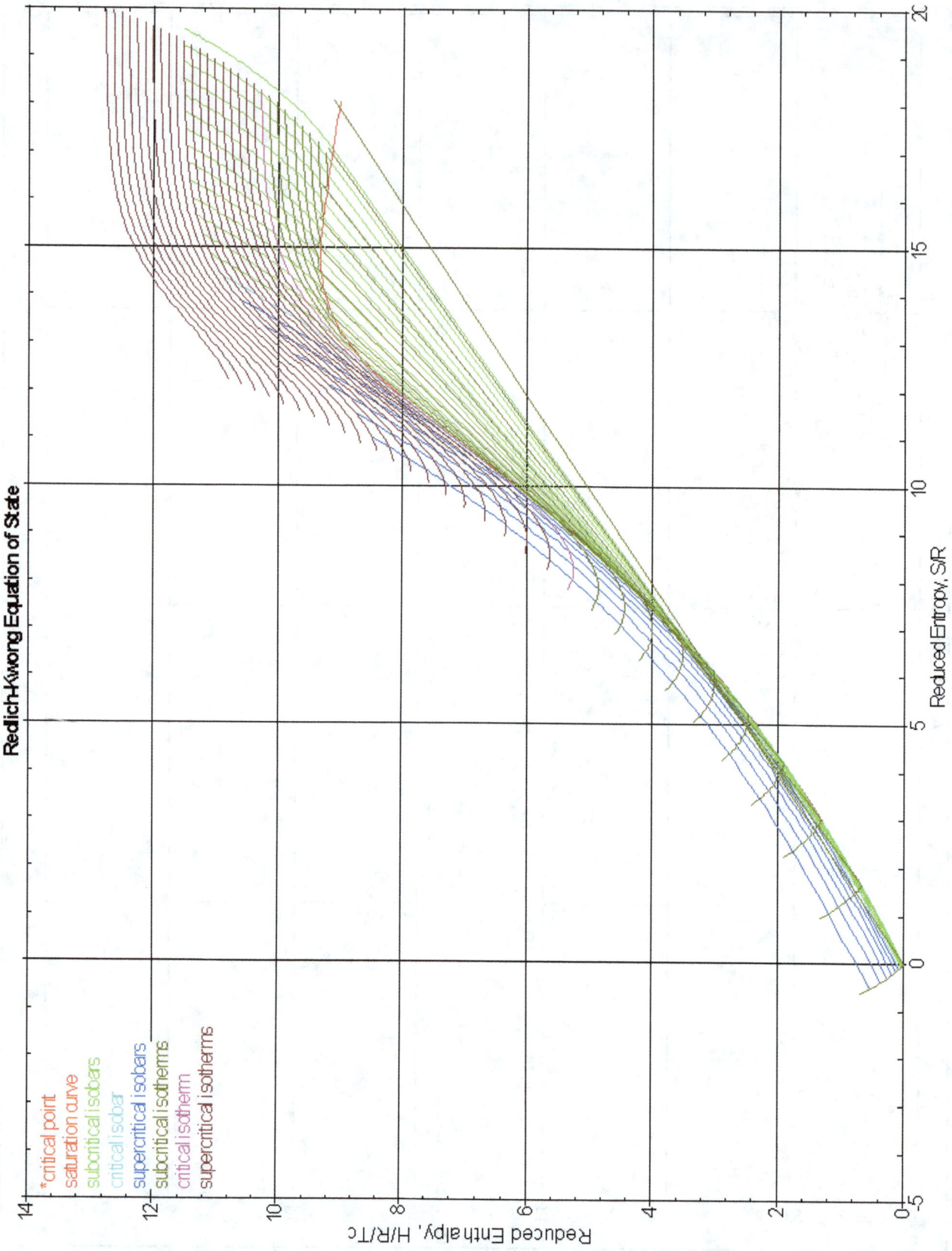

Figure 140. Mollier Chart Based on Redlich-Kwong

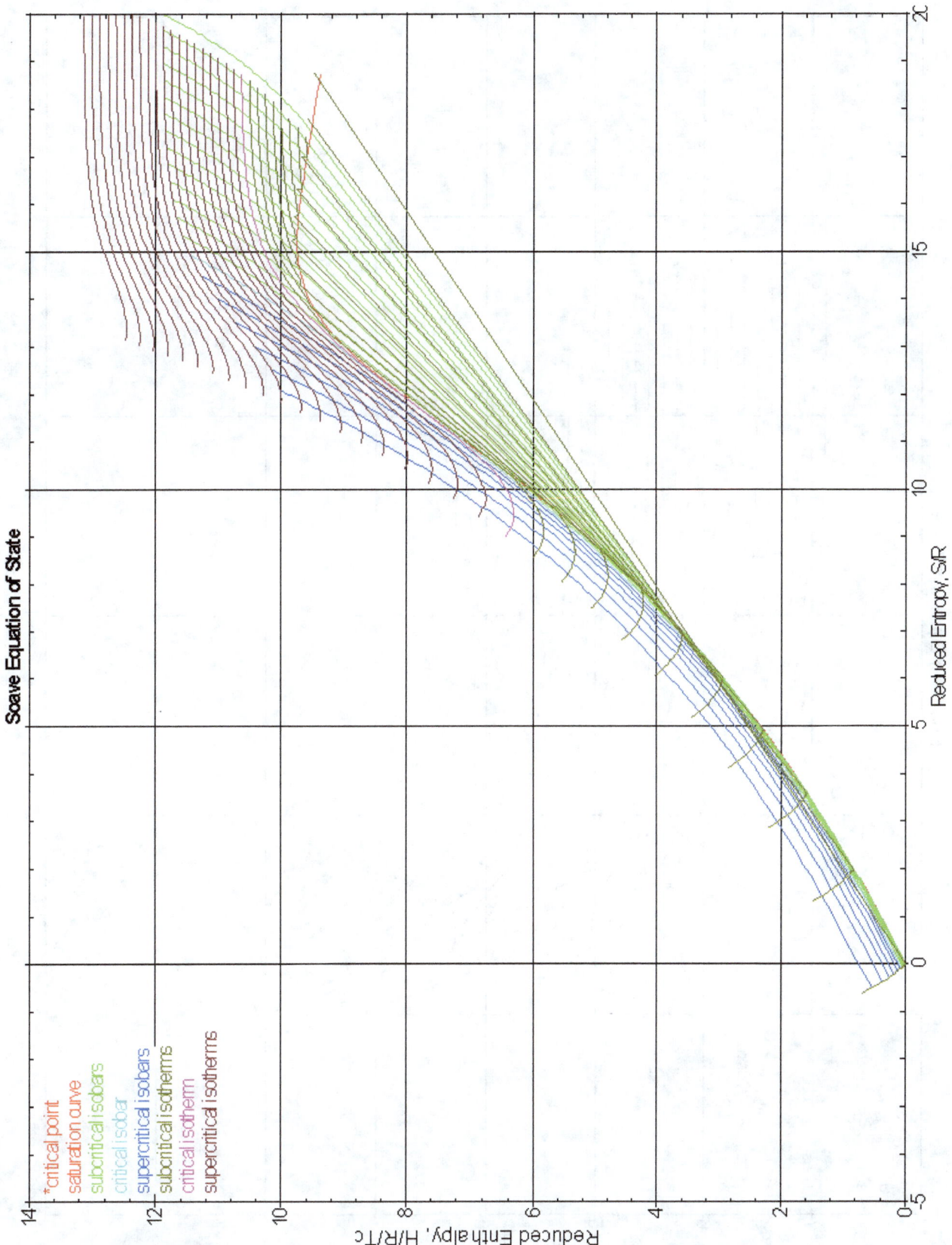

Figure 141. Mollier Chart Based on Soave's Modification

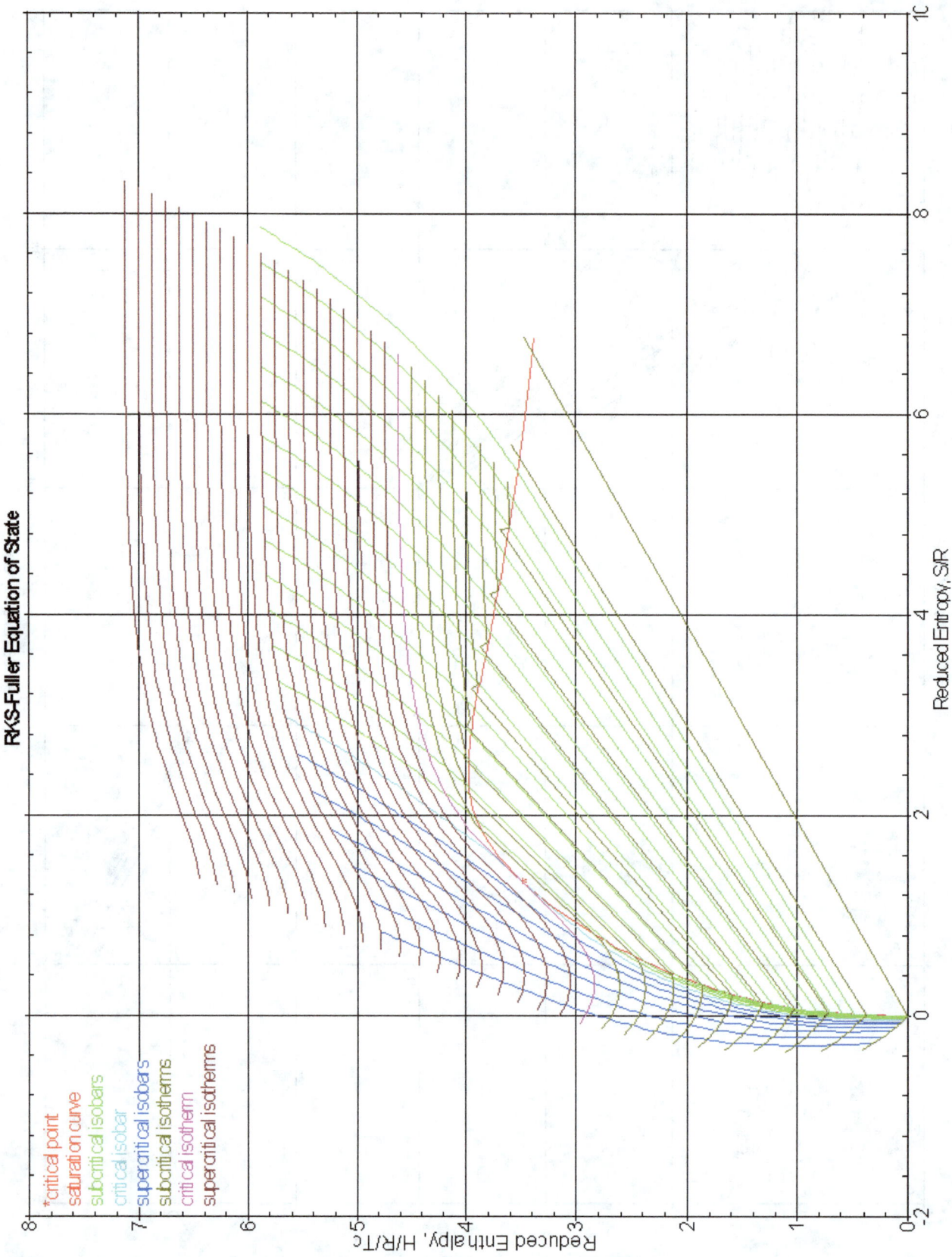

Figure 142. Mollier Chart Based on Fuller's Modification

The behavior illustrated in the figure above is not acceptable.

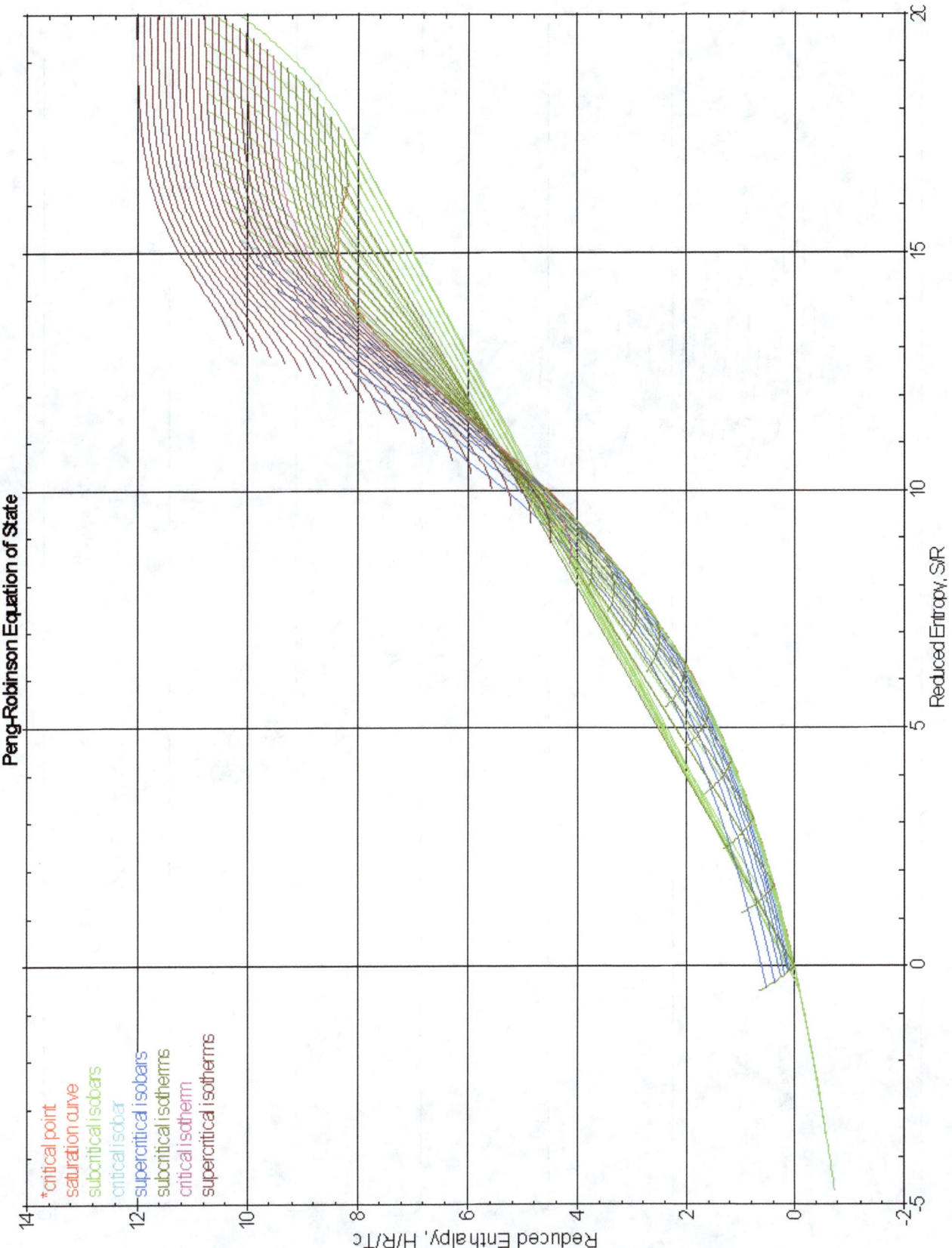

Figure 143. Mollier Chart Based on Peng-Robinson

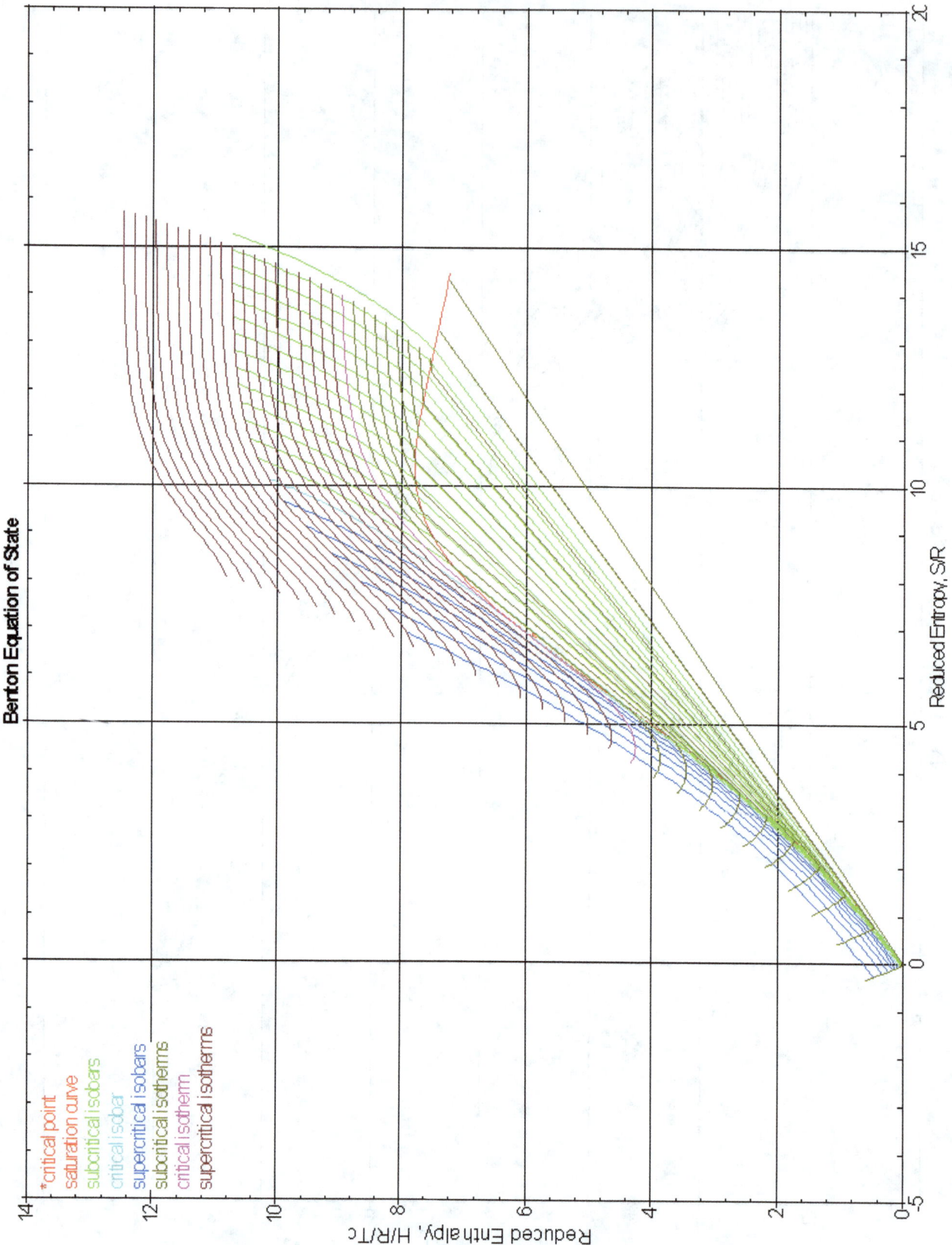

Figure 144. Mollier Chart Based on Author's Modification

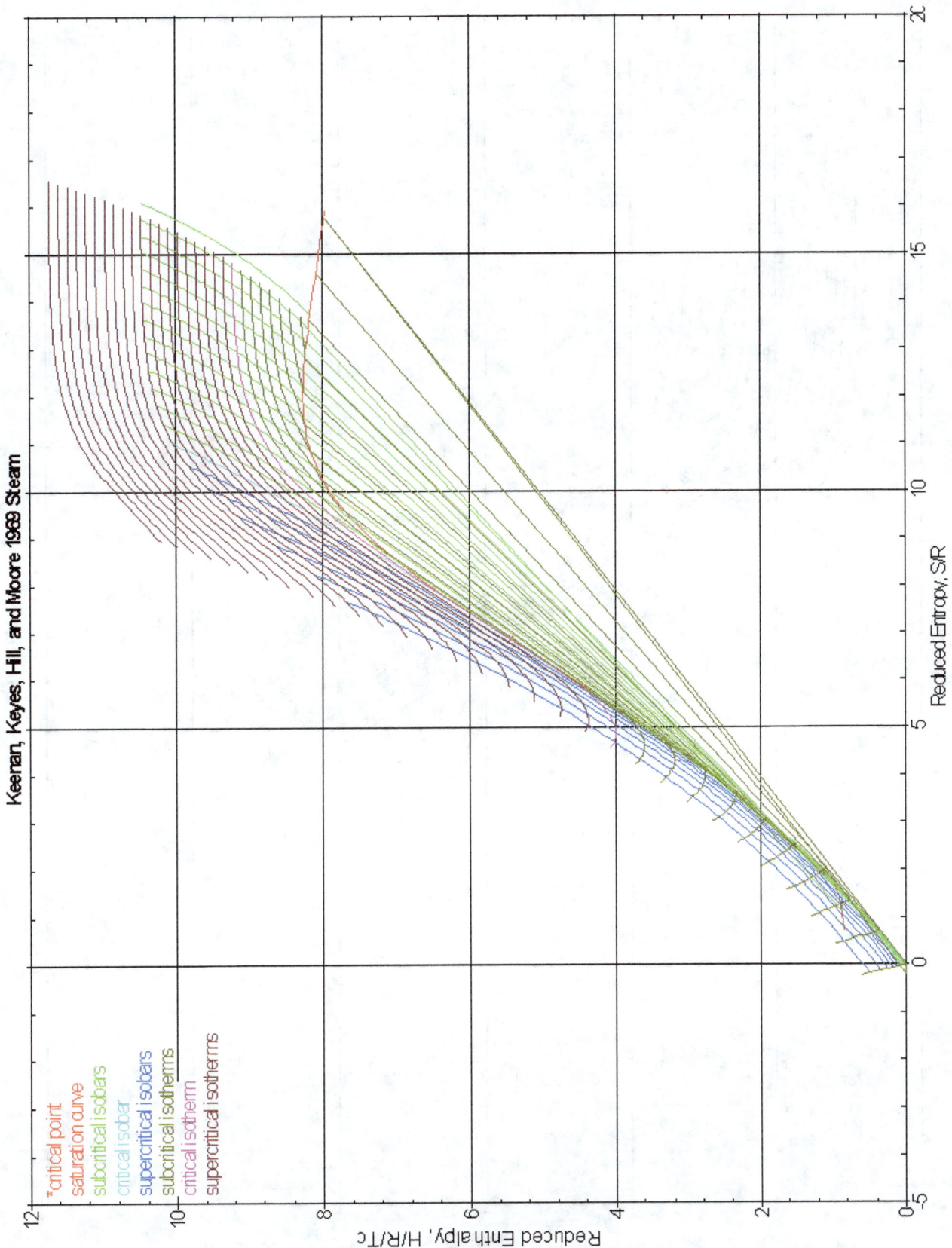

Figure 145. Mollier Chart Based on Keenan, Keyes, Hill, and Moore

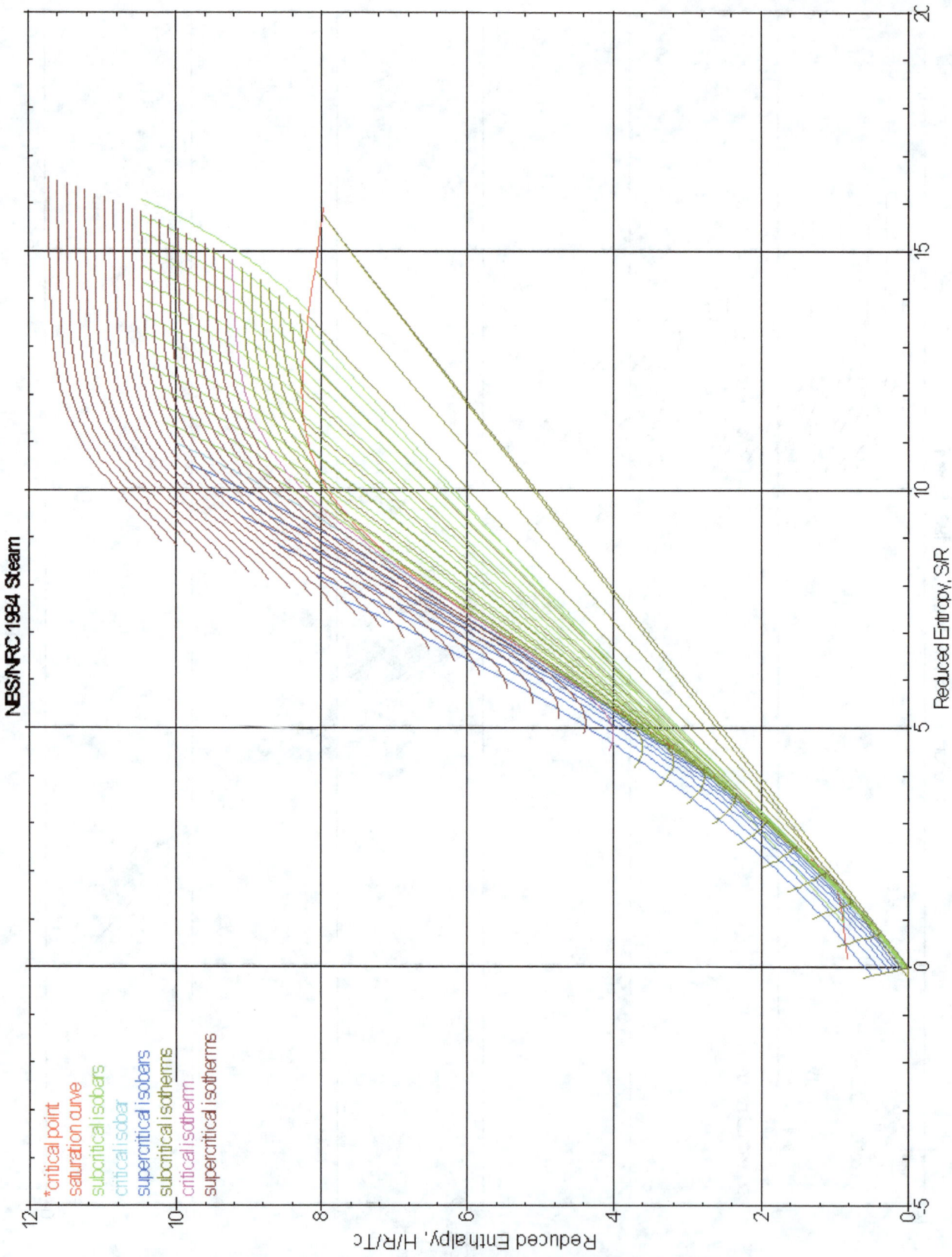

Figure 146. Mollier Chart Based on Haar, Gallagher, and Kell

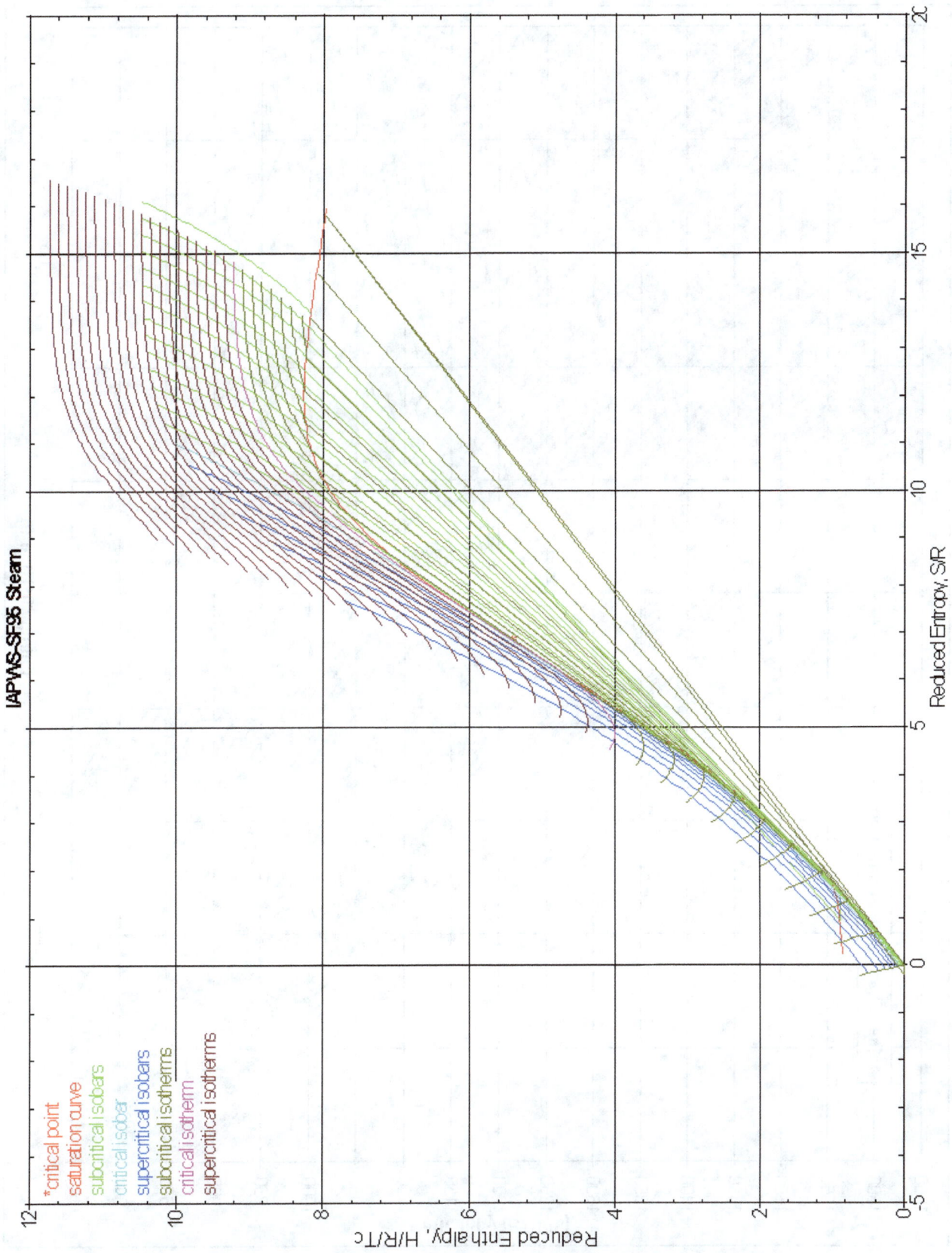

Figure 147. Mollier Chart Based on Wagner and Pruß

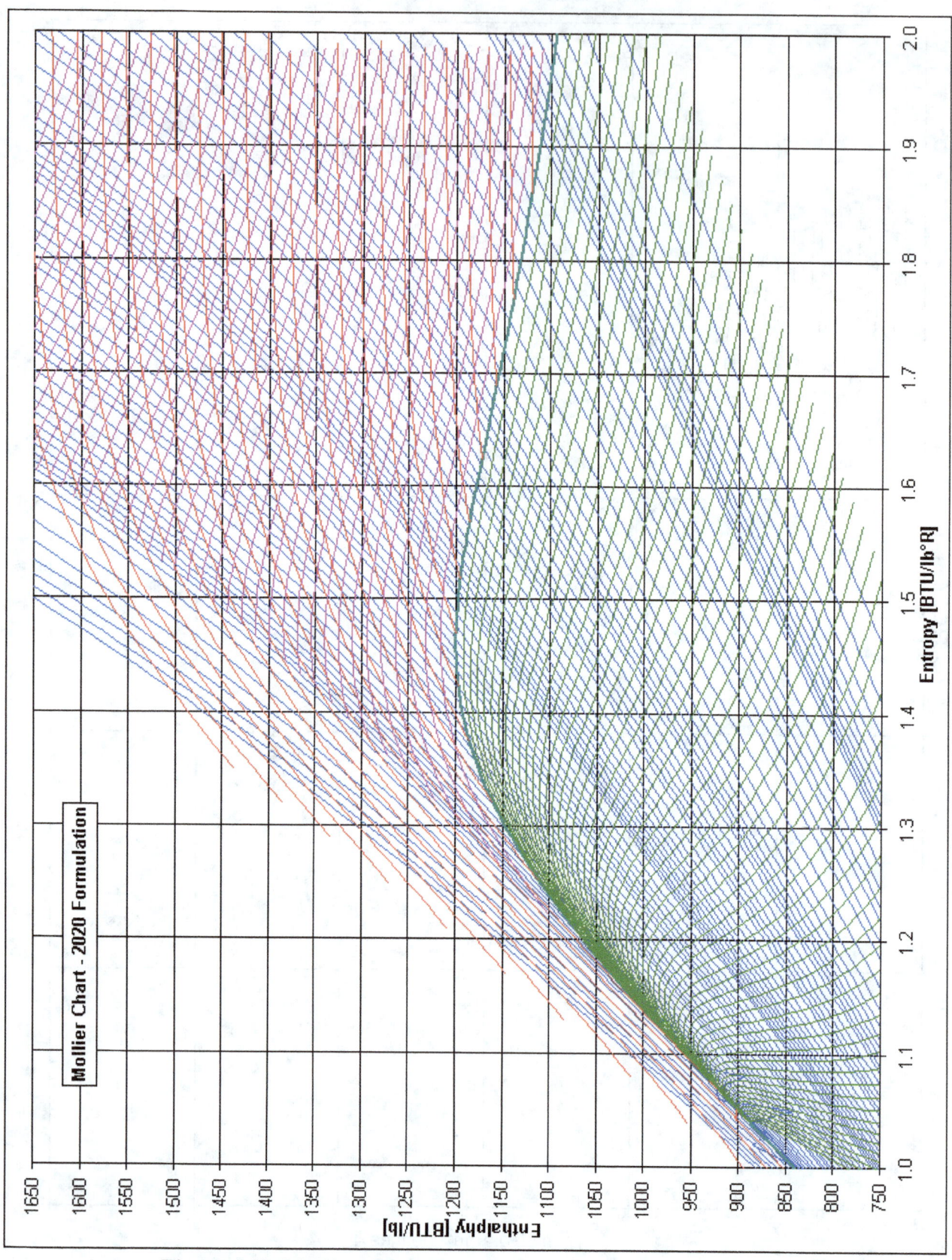

Figure 148. Mollier Chart Based on Steam 2020 Formulation

Chapter 11. ZT_R vs. $1/V_R$

The remarkable thing about this figure is that all of the isotherms are smooth and continuous, not only in value but also in slope so that it provides a unique check of the metastable region, which we cannot directly measure. In order for any formulation to be fully accurate over the entire range of conditions, these isotherms must be smooth and continuous. As the following figures illustrate, this condition is met by only a few of the available formulas and only one of those representing steam—the 2020.

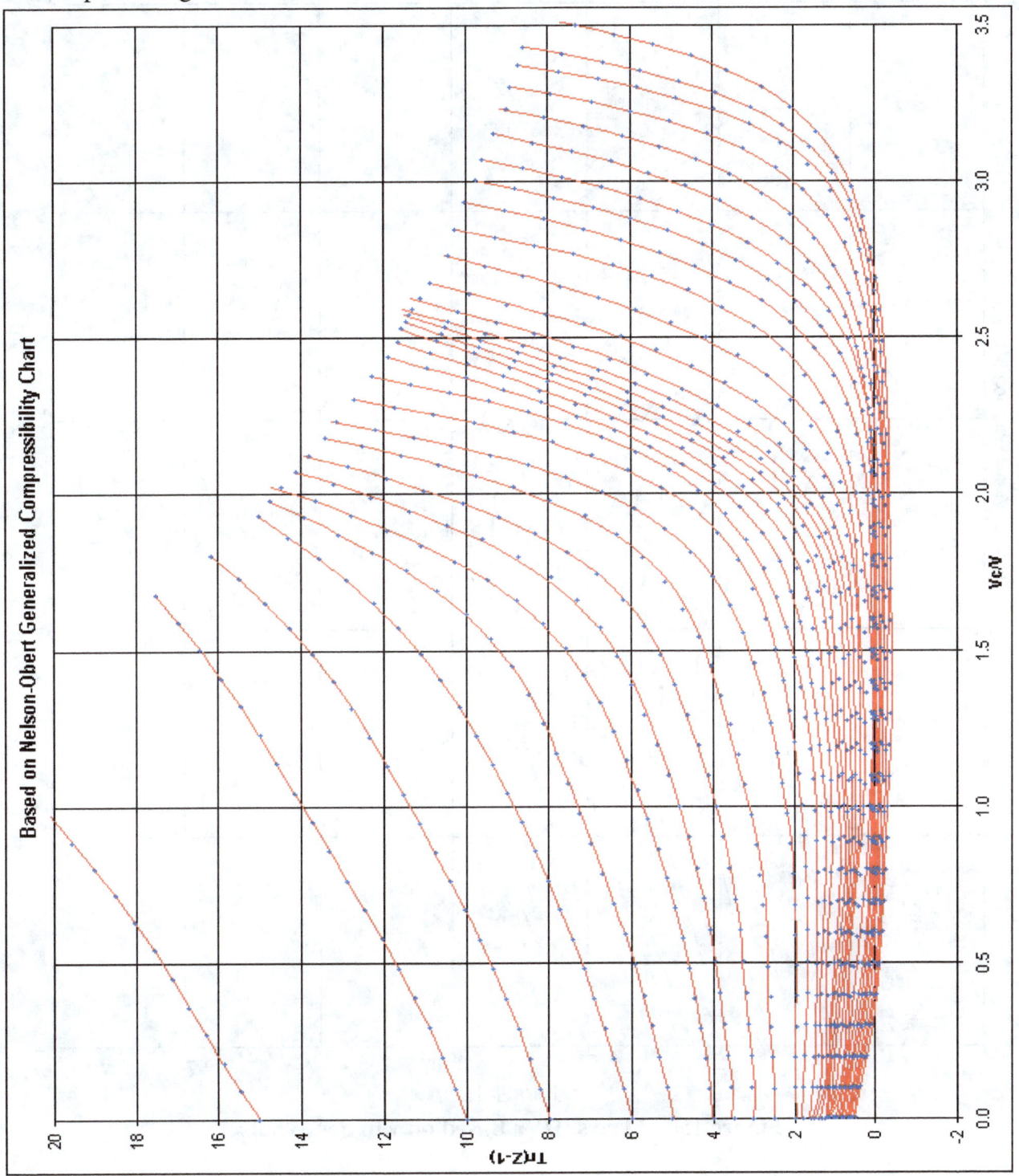

Figure 149. ZT_r vs. $1/V_r$ Based on Nelson-Obert

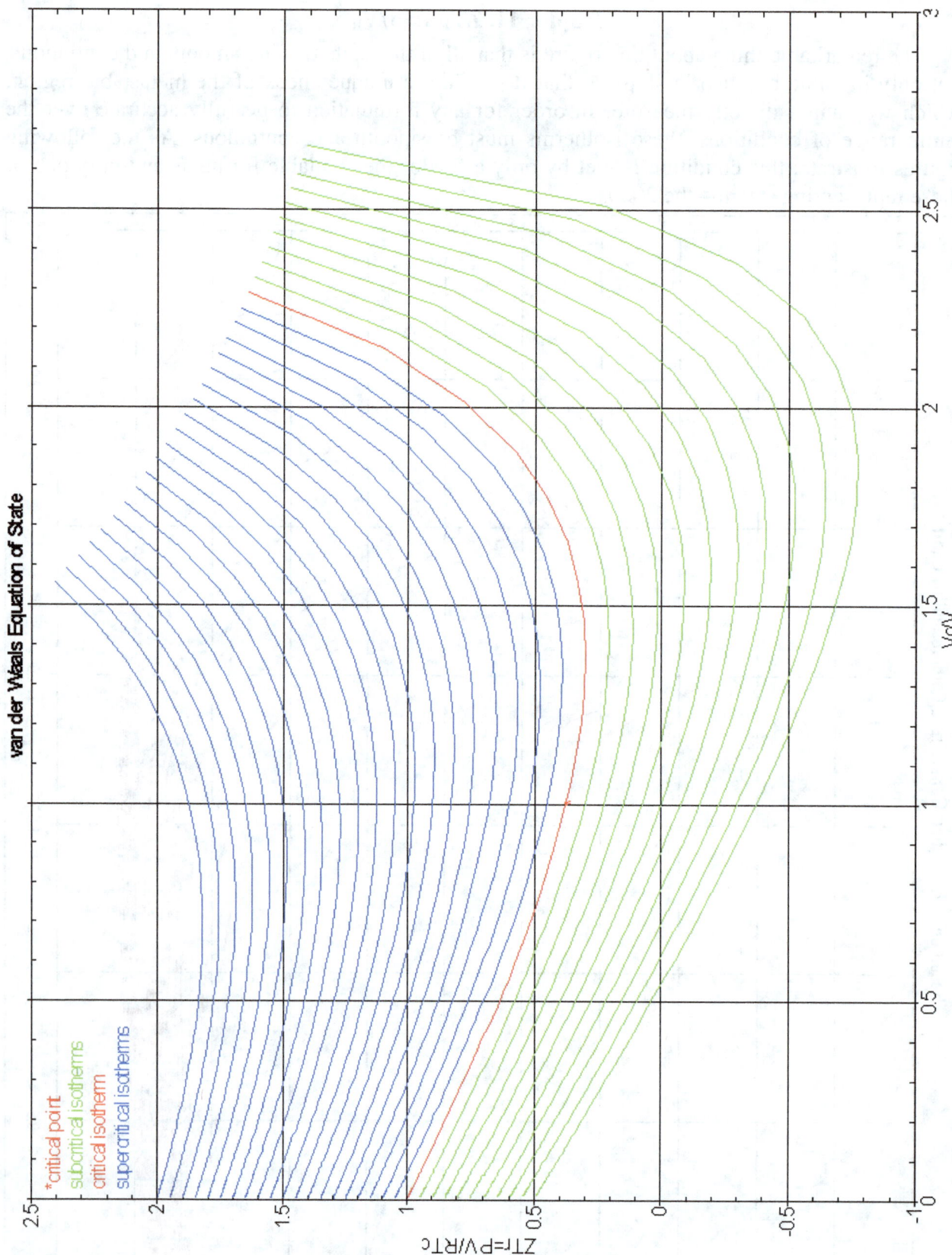

Figure 150. ZTr vs. 1/Vr Based on van der Waals

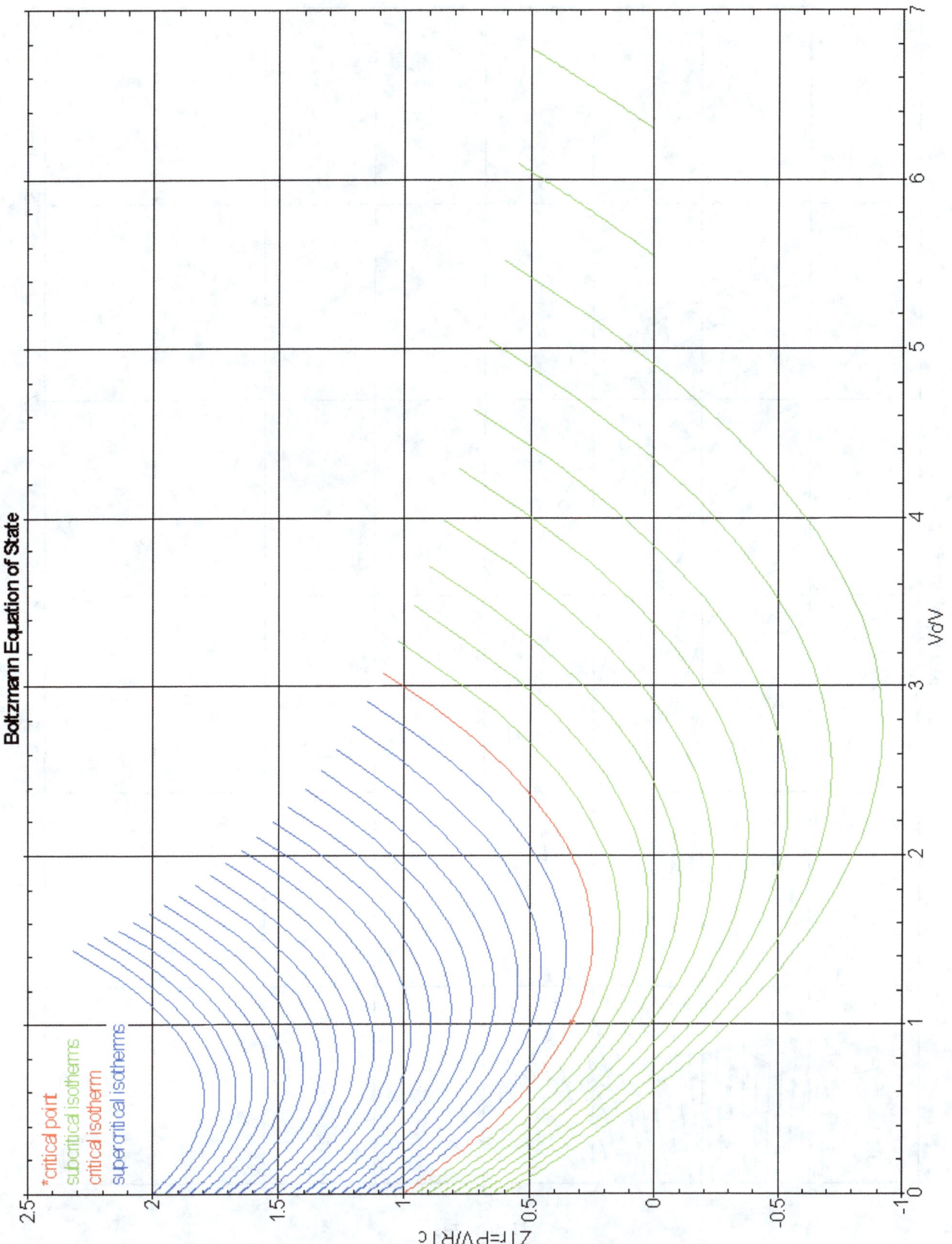

Figure 151. ZTr vs. 1/Vr Based on Boltzmann

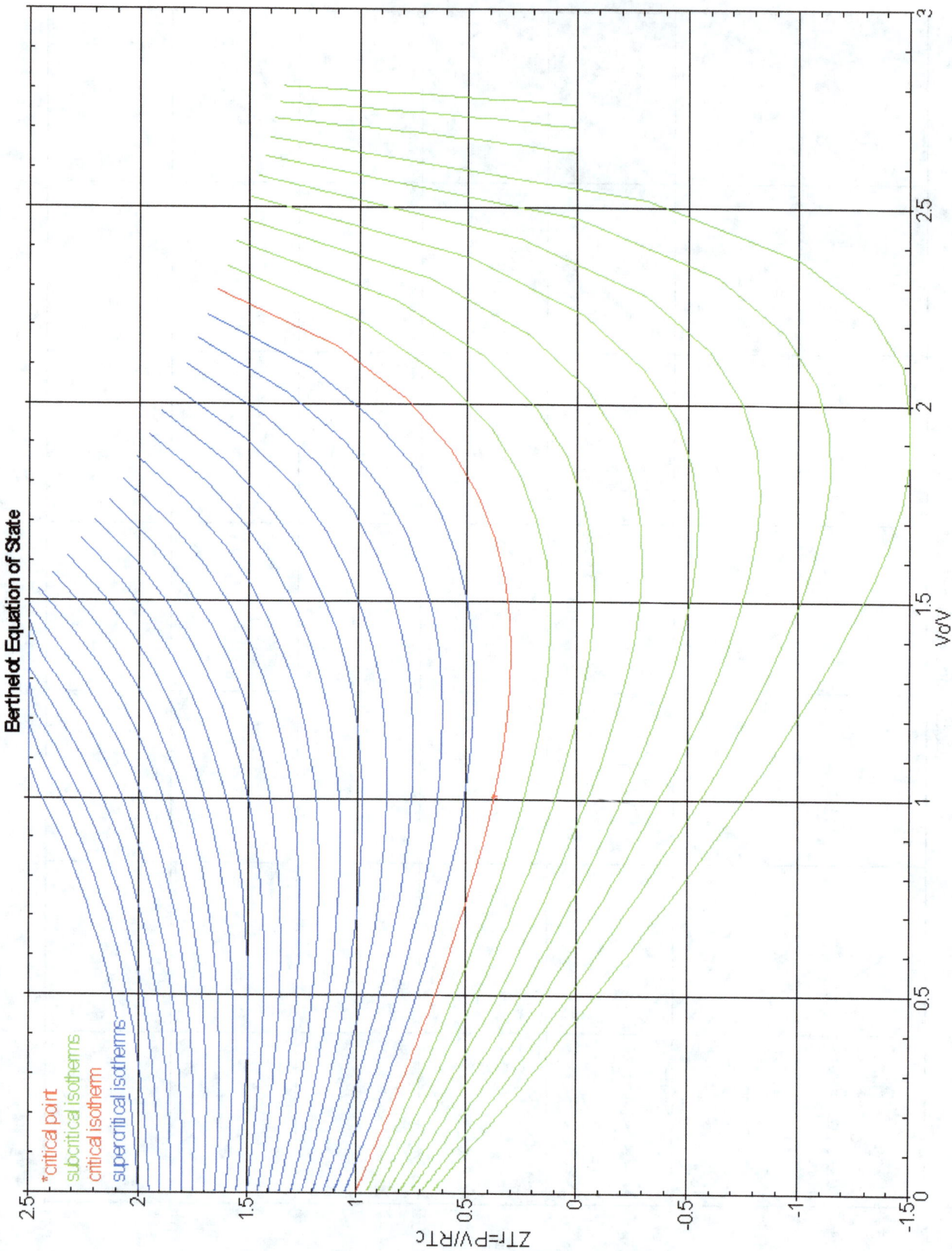

Figure 152. ZTr vs. 1/Vr Based on Berthelot

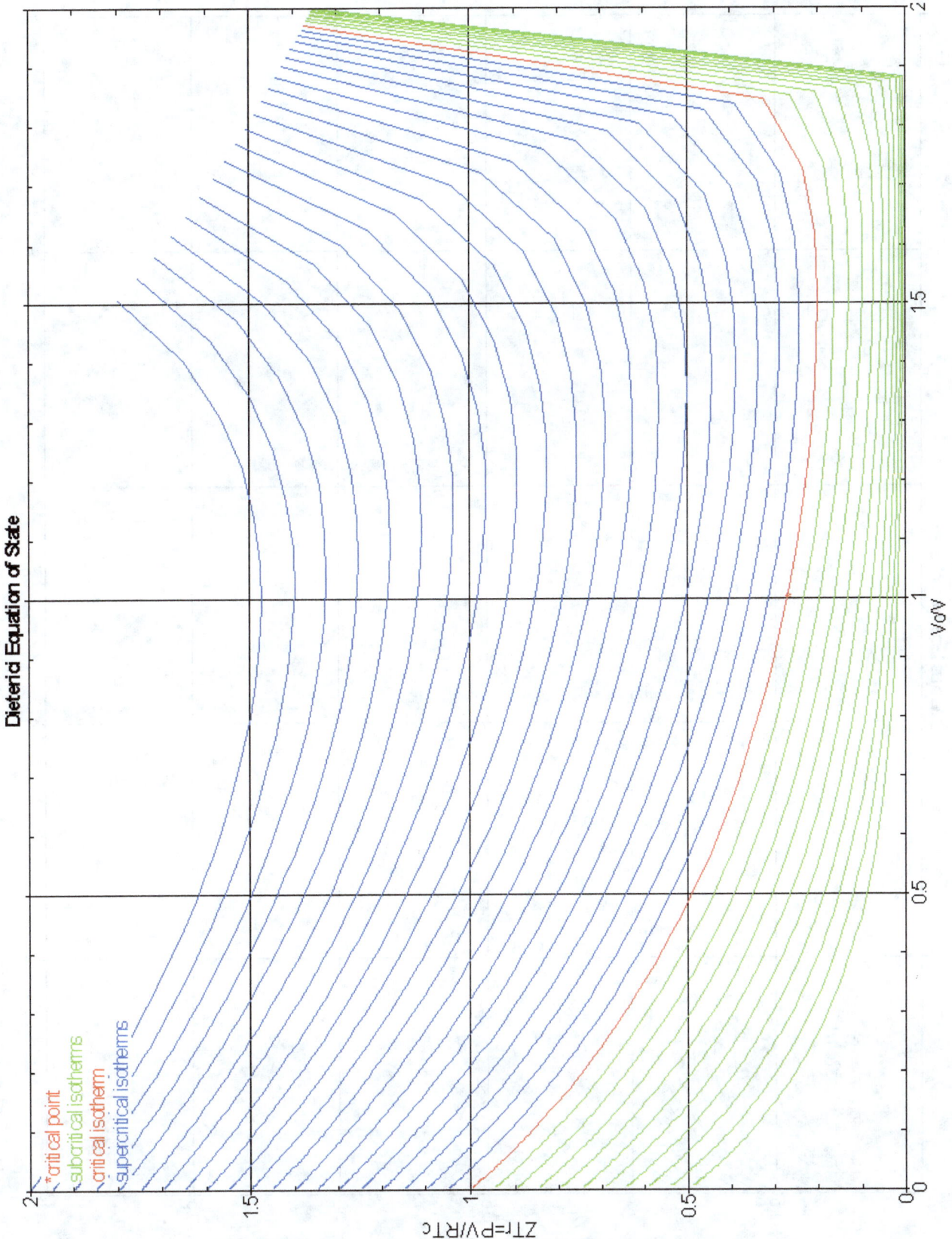

Figure 153. ZTr vs. 1/Vr Based on Dieterici

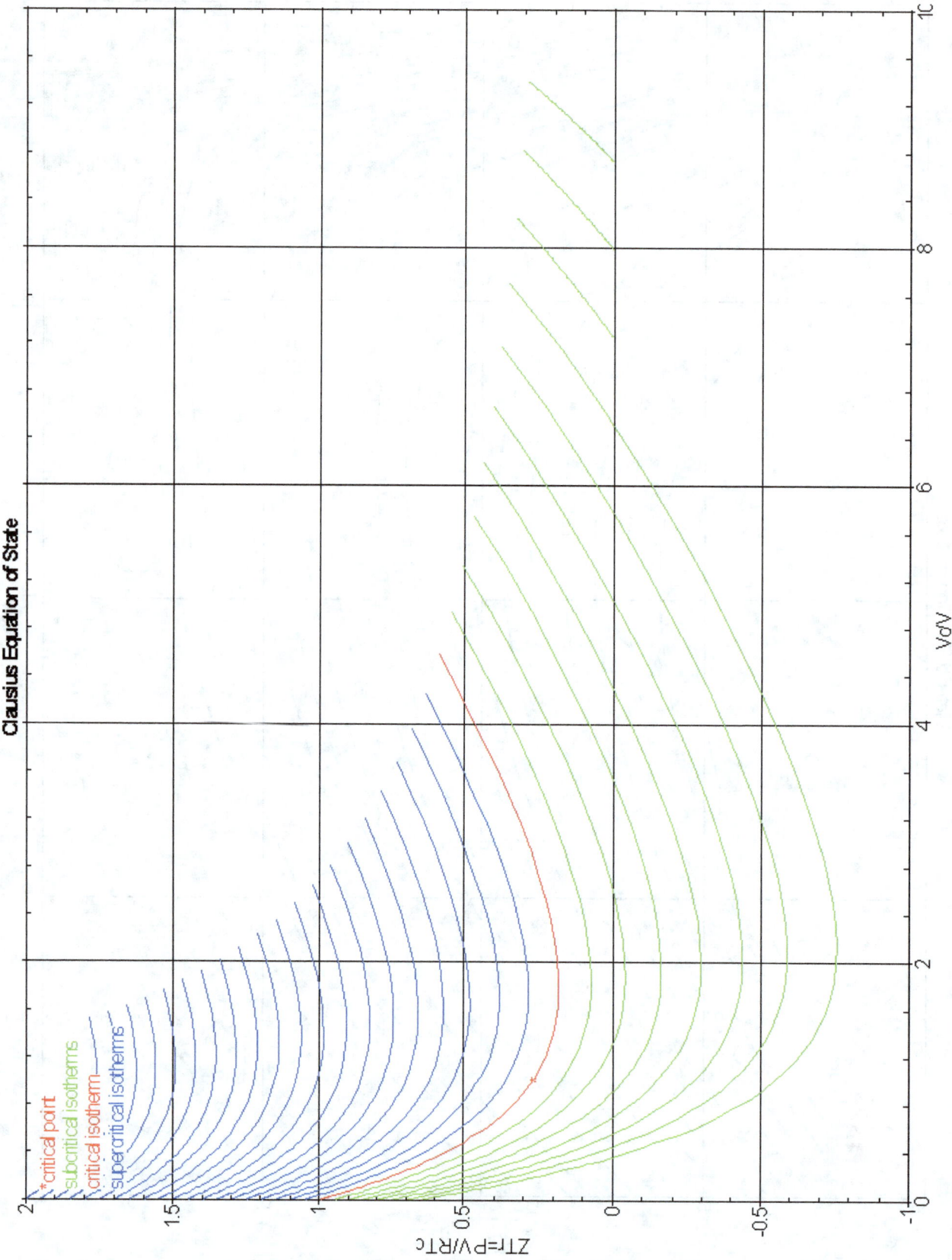

Figure 154. ZTr vs. 1/Vr Based on Clausius

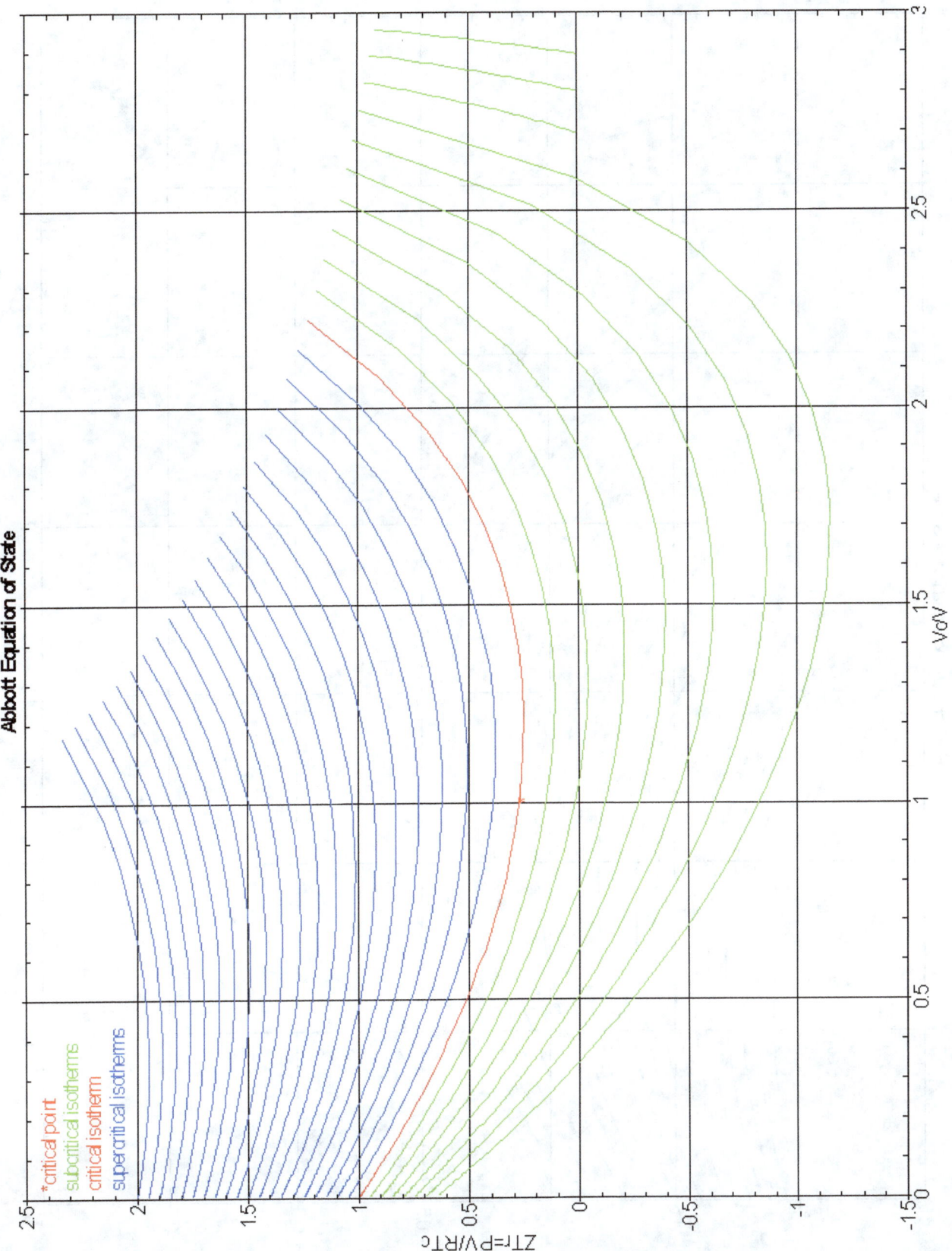

Figure 155. ZTr vs. 1/Vr Based on Abbott's Modification

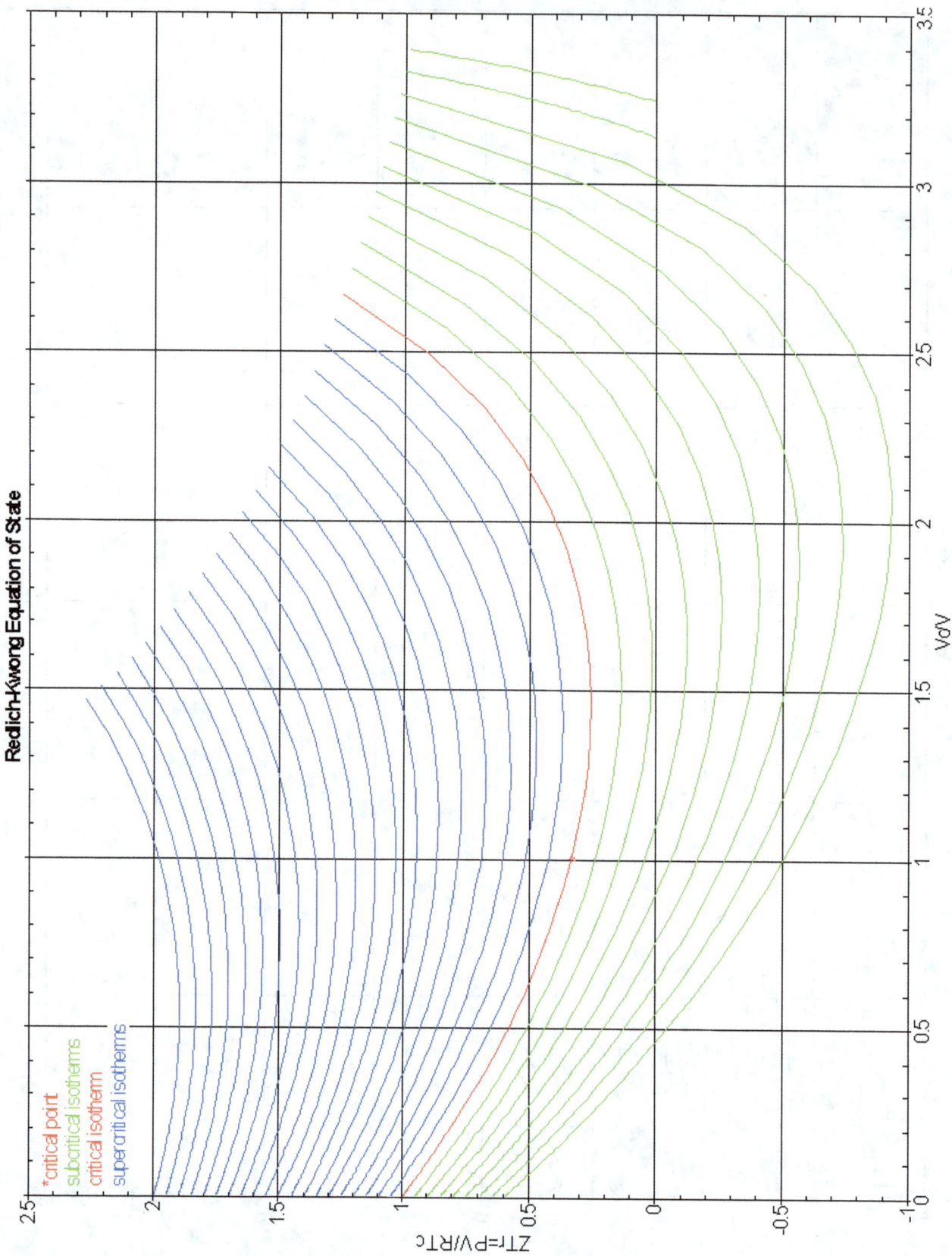

Figure 156. ZTr vs. 1/Vr Based on Redlich-Kwong

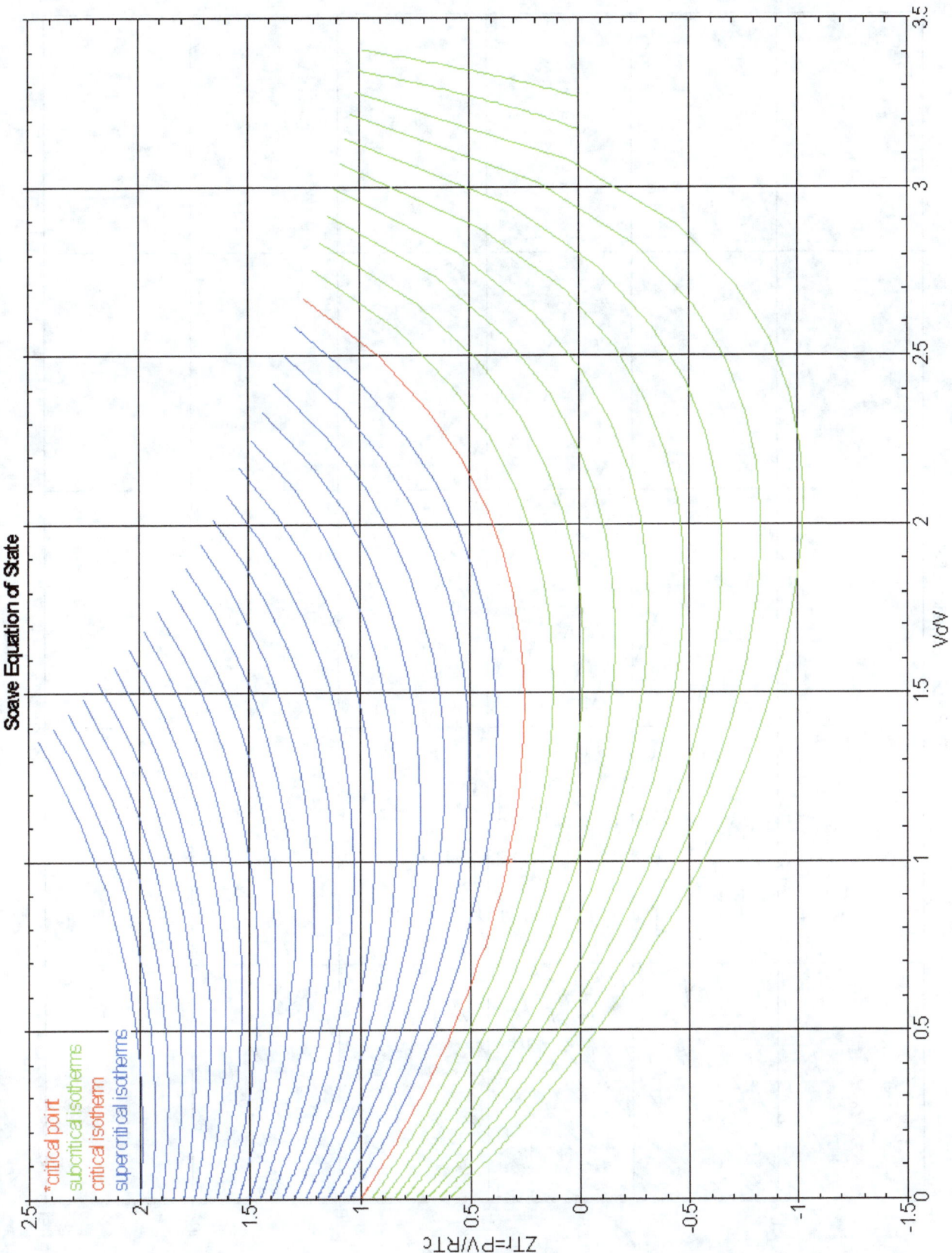

Figure 157. ZTr vs. 1/Vr Based on Soave's Modification

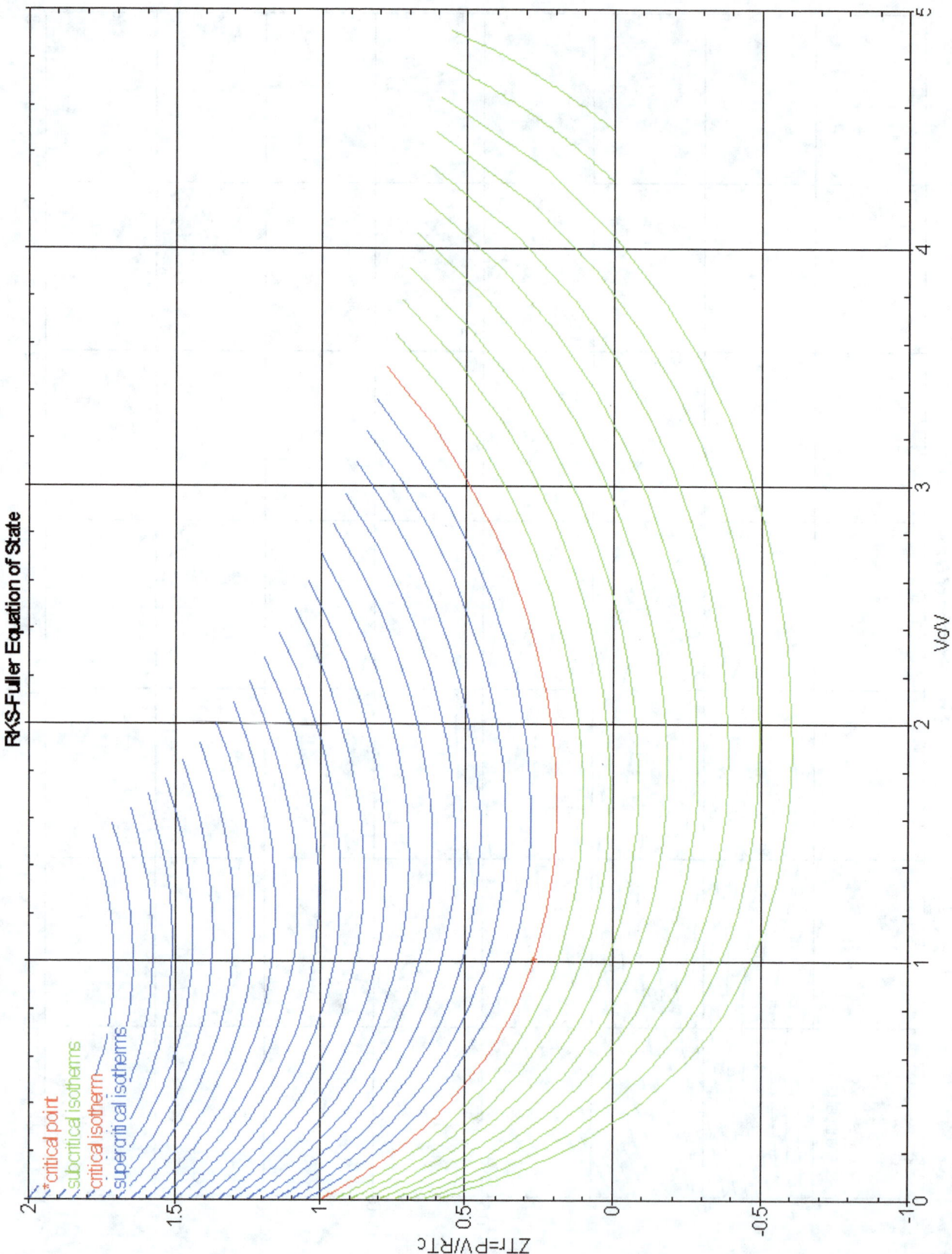

Figure 158. ZTr vs. 1/Vr Based on Fuller's Modification

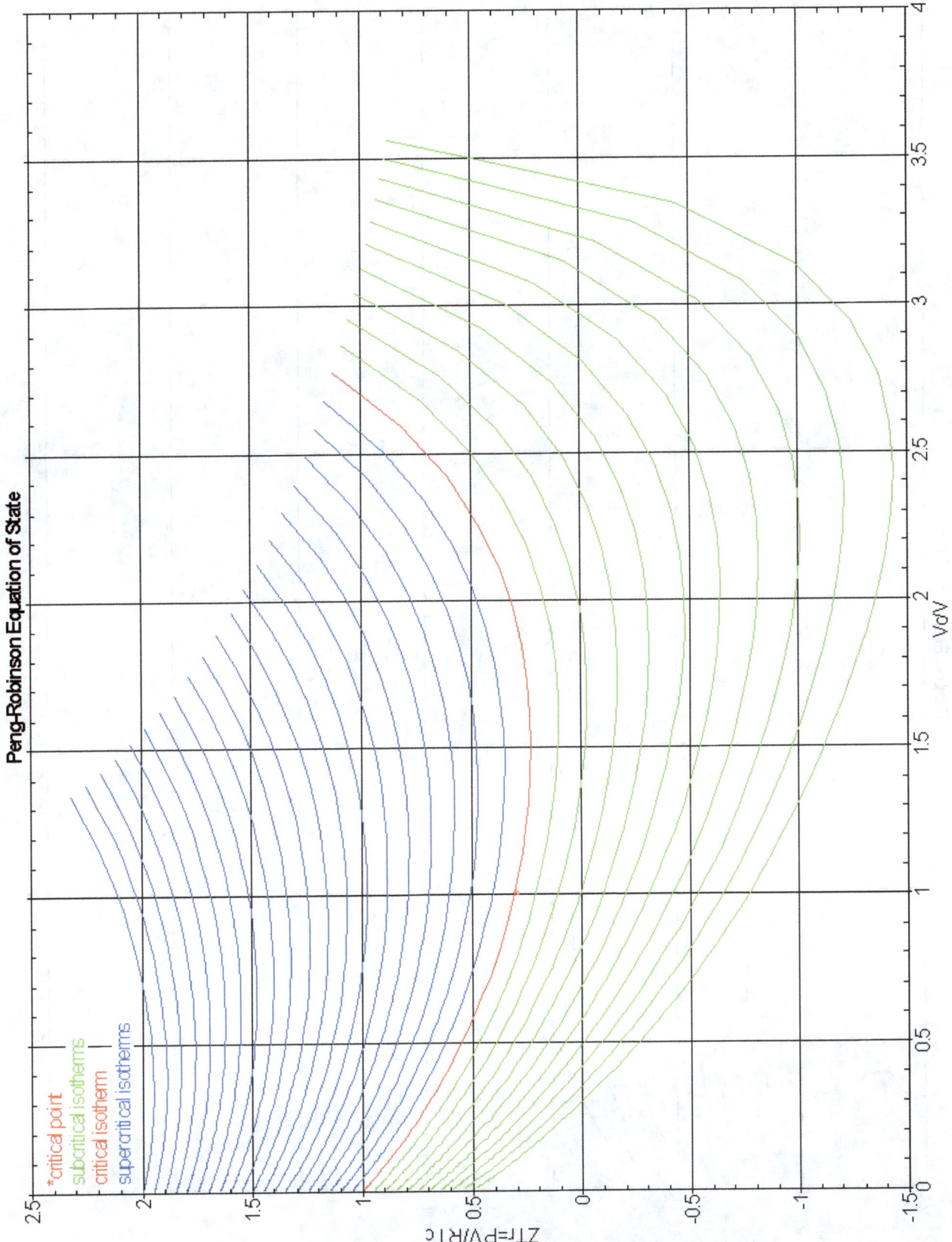

Figure 159. ZTr vs. 1/Vr Based on Peng-Robinson

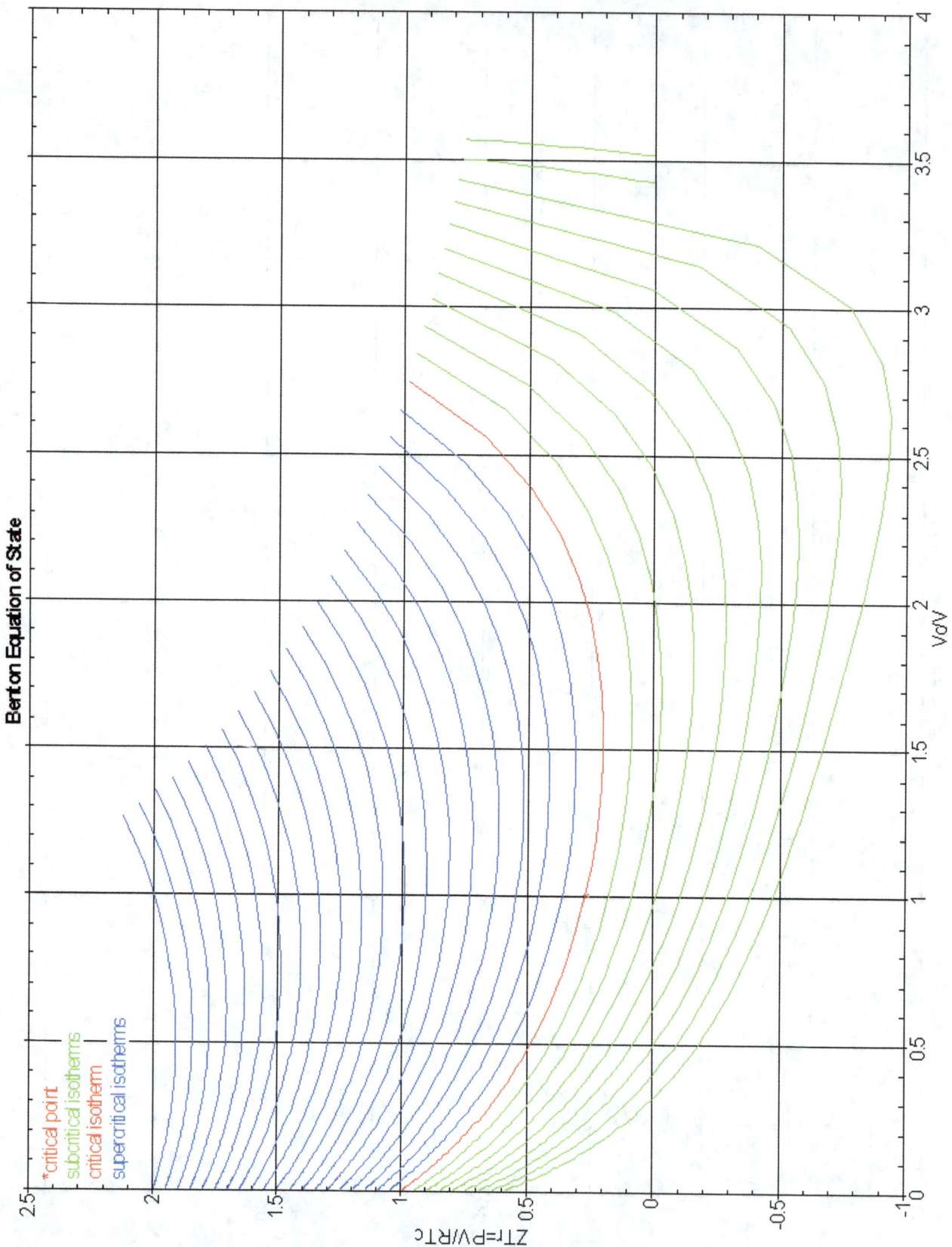

Figure 160. ZTr vs. 1/Vr Based on Author's Modification

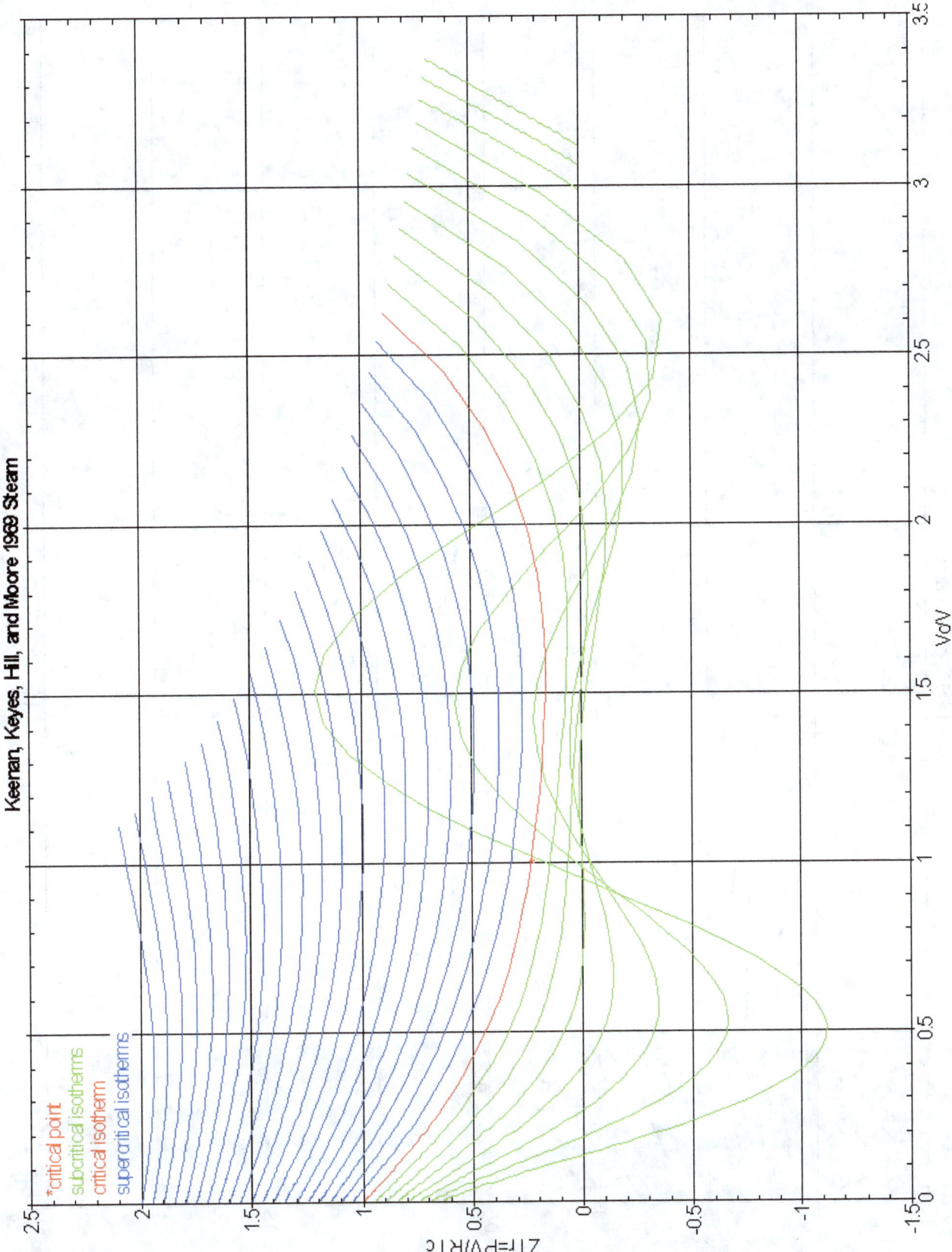

Figure 161. ZTr vs. 1/Vr Based on Keenan, Keyes, Hill, and Moore

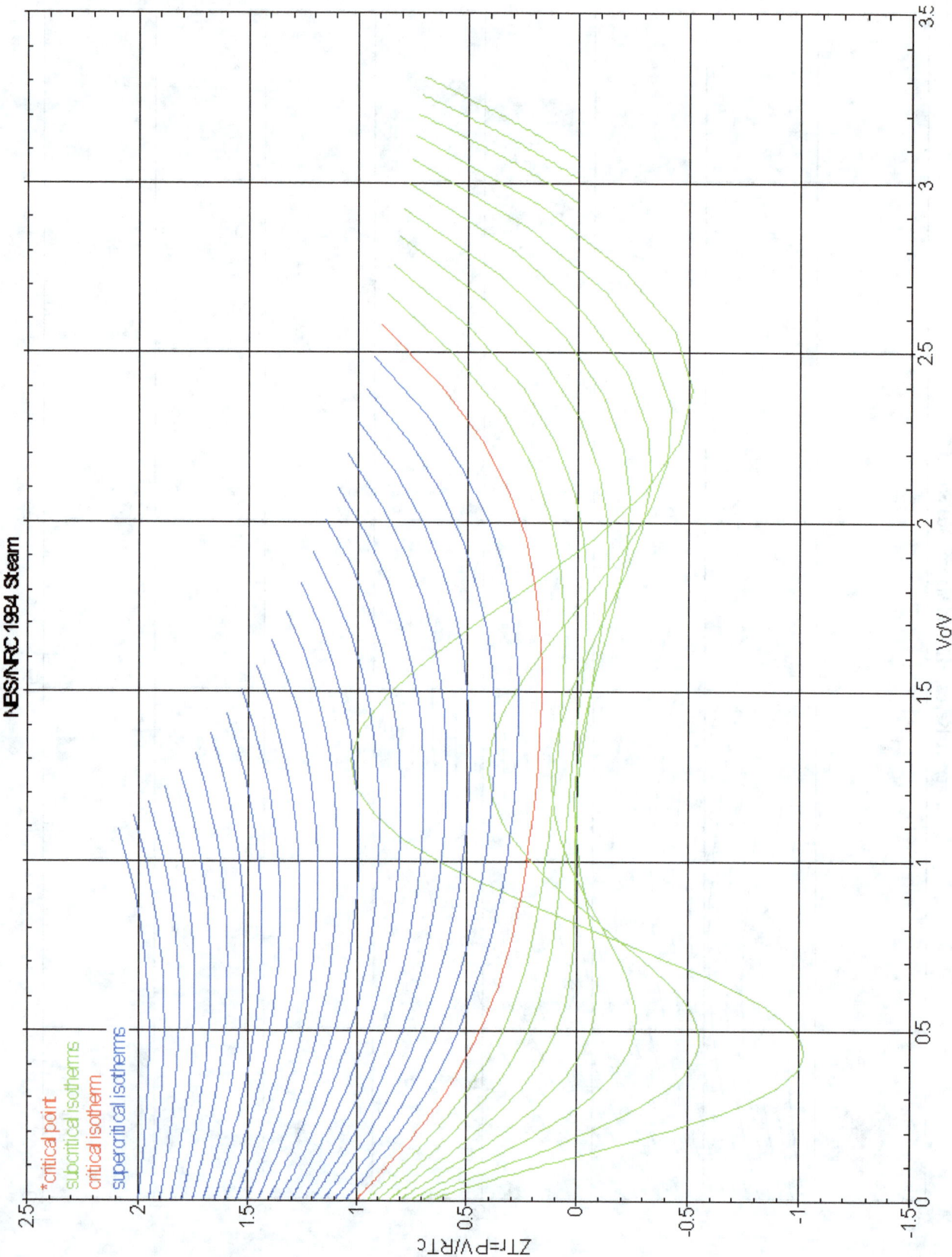

Figure 162. ZTr vs. 1/Vr Based on Haar, Gallagher, and Kell

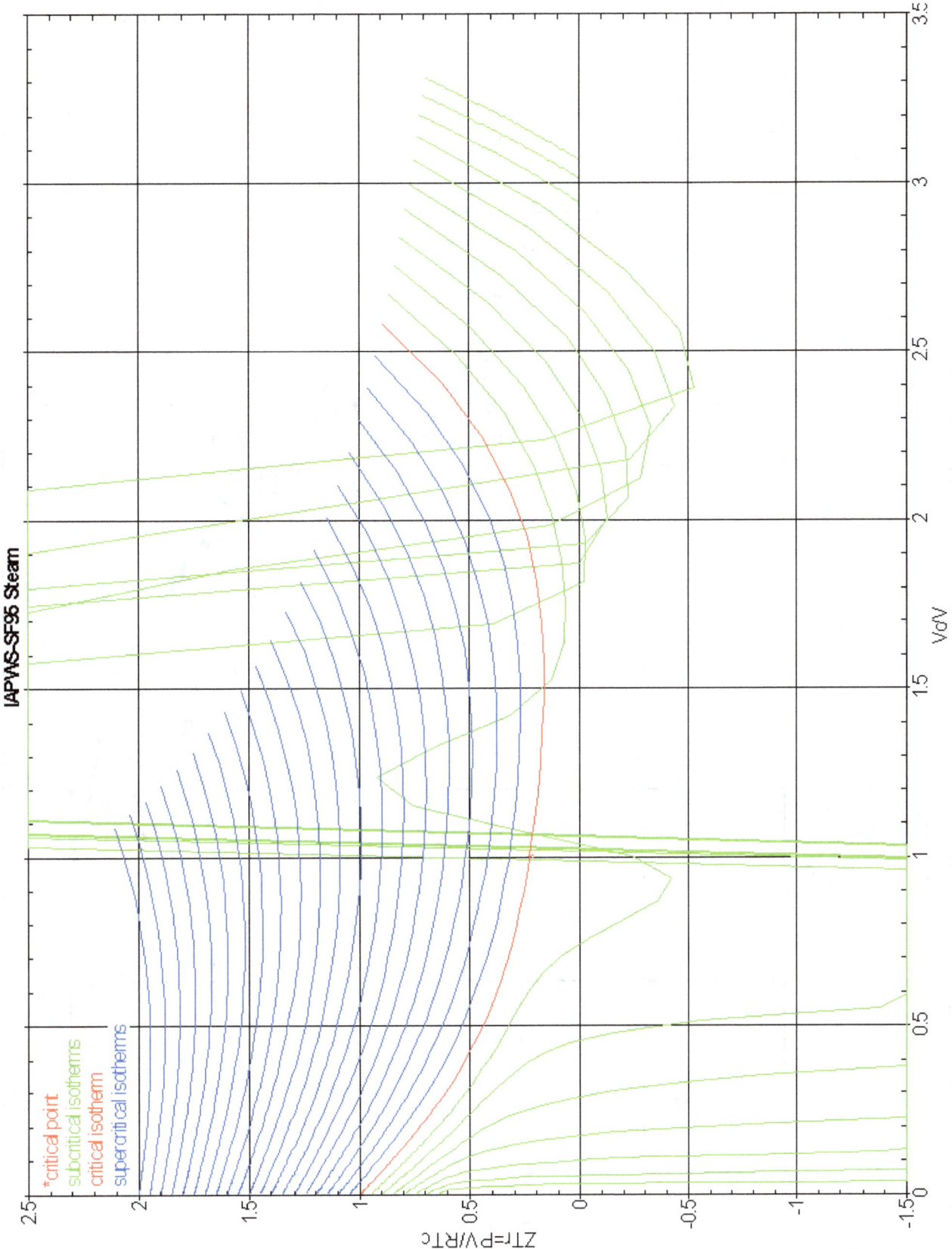

Figure 163. ZTr vs. 1/Vr Based on Wagner and Pruß

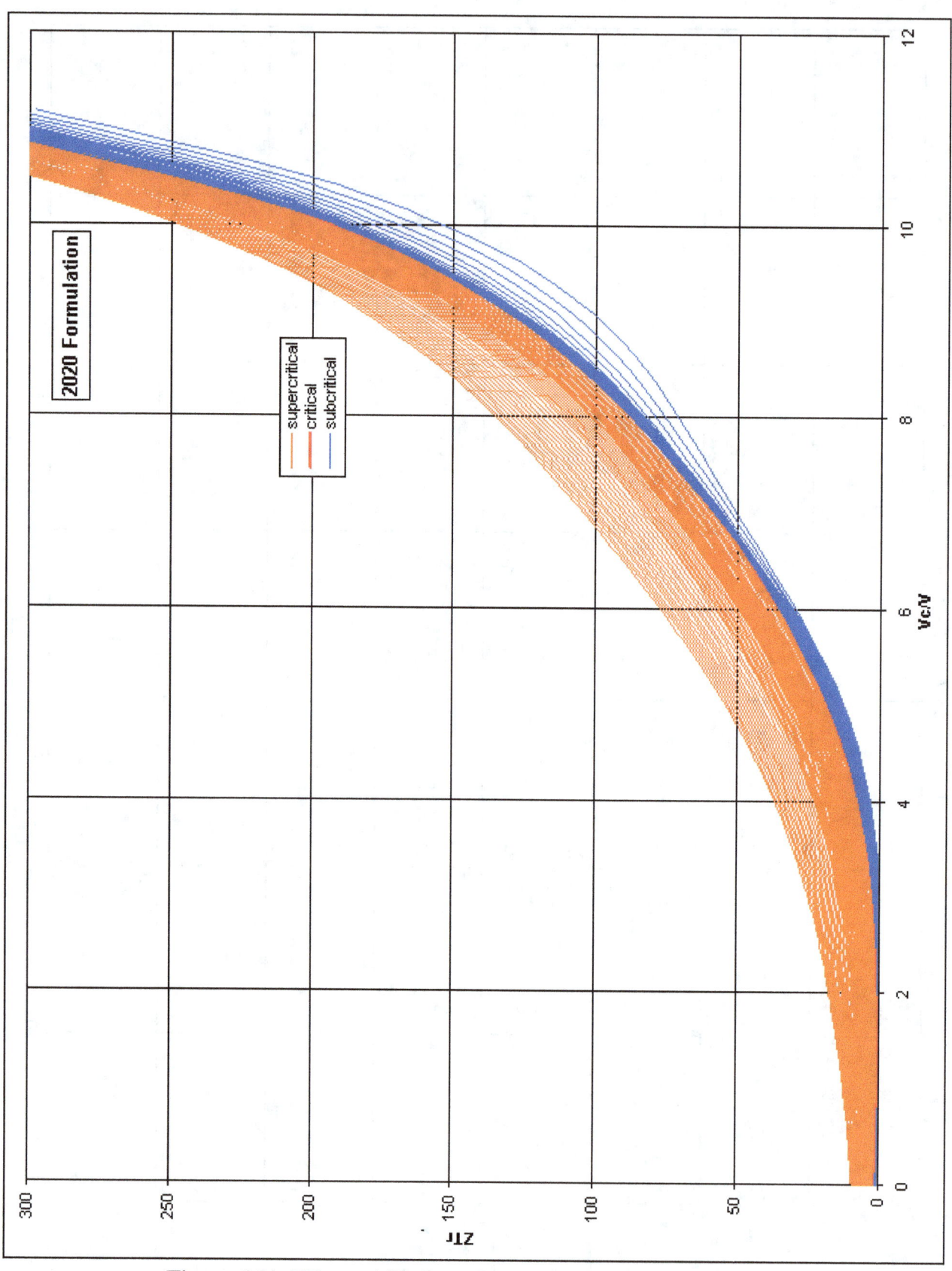

Figure 164. ZTr vs. 1/Vr Based on Steam 2020 Formulation

Note that this figure covers a much more extensive range than the others in this chapter.

also by D. James Benton

3D Articulation: Using OpenGL, ISBN-9798596362480, Amazon, 2021 (book 3 in the 3D series).

3D Models in Motion Using OpenGL, ISBN-9798652987701, Amazon, 2020 (book 2 in the 3D series.

3D Rendering in Windows: How to display three-dimensional objects in Windows with and without OpenGL, ISBN-9781520339610, Amazon, 2016 (book 1 in the 3D series).

A Synergy of Short Stories: The whole may be greater than the sum of the parts, ISBN-9781520340319, Amazon, 2016.

Azeotropes: Behavior and Application, ISBN-9798609748997, Amazon, 2020.

bat-Elohim: Book 3 in the Little Star Trilogy, ISBN-9781686148682, Amazon, 2019.

Boilers: Performance and Testing, ISBN: 9798789062517, Amazon 2021.

Combined 3D Rendering Series: 3D Rendering in Windows®, 3D Models in Motion, and 3D Articulation, ISBN-9798484417032, Amazon, 2021.

Complex Variables: Practical Applications, ISBN-9781794250437, Amazon, 2019.

Compression & Encryption: Algorithms & Software, ISBN-9781081008826, Amazon, 2019.

Computational Fluid Dynamics: an Overview of Methods, ISBN-9781672393775, Amazon, 2019.

Computer Simulation of Power Systems: Programming Strategies and Practical Examples, ISBN-9781696218184, Amazon, 2019.

Contaminant Transport: A Numerical Approach, ISBN-9798461733216, Amazon, 2021.

CPUnleashed! Tapping Processor Speed, ISBN-9798421420361, Amazon, 2022.

Curve-Fitting: The Science and Art of Approximation, ISBN-9781520339542, Amazon, 2016.

Death by Tie: It was the best of ties. It was the worst of ties. It's what got him killed., ISBN-9798398745931, Amazon, 2023.

Differential Equations: Numerical Methods for Solving, ISBN-9781983004162, Amazon, 2018.

Evaporative Cooling: The Science of Beating the Heat, ISBN-9781520913346, Amazon, 2017.

Forecasting: Extrapolation and Projection, ISBN-9798394019494, Amazon 2023.

Heat Engines: Thermodynamics, Cycles, & Performance Curves, ISBN-9798486886836, Amazon, 2021.

Heat Exchangers: Performance Prediction & Evaluation, ISBN-9781973589327, Amazon, 2017.

Heat Recovery Steam Generators: Thermal Design and Testing, ISBN-9781691029365, Amazon, 2019.

Heat Transfer: Heat Exchangers, Heat Recovery Steam Generators, & Cooling Towers, ISBN-9798487417831, Amazon, 2021.

Heat Transfer Examples: Practical Problems Solved, ISBN-9798390610763, Amazon, 2023.

The Kick-Start Murders: Visualize revenge, ISBN-9798759083375, Amazon, 2021.

Jamie2: Innocence is easily lost and cannot be restored, ISBN-9781520339375, Amazon, 2016-18.

Kyle Cooper Mysteries: Kick Start, Monte Carlo, and Waterfront Murders, ISBN-9798829365943, Amazon, 2022.

The Last Seraph: Sequel to Little Star, ISBN-9781726802253, Amazon, 2018.

Little Star: God doesn't do things the way we expect Him to. He's better than that! ISBN-9781520338903, Amazon, 2015-17.

Living Math: Seeing mathematics in every day life (and appreciating it more too), ISBN-9781520336992, Amazon, 2016.

Lost Cause: If only history could be changed…, ISBN-9781521173770, Amazon, 2017.

Mass Transfer: Diffusion & Convection, ISBN-9798702403106, Amazon, 2021.

Mill Town Destiny: The Hand of Providence brought them together to rescue the mill, the town, and each other, ISBN-9781520864679, Amazon, 2017.
Monte Carlo Murders: Who Killed Who and Why, ISBN-9798829341848, Amazon, 2022.
Monte Carlo Simulation: The Art of Random Process Characterization, ISBN-9781980577874, Amazon, 2018.
Nonlinear Equations: Numerical Methods for Solving, ISBN-9781717767318, Amazon, 2018.
Numerical Calculus: Differentiation and Integration, ISBN-9781980680901, Amazon, 2018.
Numerical Methods: Nonlinear Equations, Numerical Calculus, & Differential Equations, ISBN-9798486246845, Amazon, 2021.
Orthogonal Functions: The Many Uses of, ISBN-9781719876162, Amazon, 2018.
Overwhelming Evidence: A Pilgrimage, ISBN-9798515642211, Amazon, 2021.
Particle Tracking: Computational Strategies and Diverse Examples, ISBN-9781692512651, Amazon, 2019.
Plumes: Delineation & Transport, ISBN-9781702292771, Amazon, 2019.
Power Plant Performance Curves: for Testing and Dispatch, ISBN-9798640192698, Amazon, 2020.
Practical Linear Algebra: Principles & Software, ISBN-9798860910584, Amazon, 2023.
Props, Fans, & Pumps: Design & Performance, ISBN-9798645391195, Amazon, 2020.
Remediation: Contaminant Transport, Particle Tracking, & Plumes, ISBN-9798485651190, Amazon, 2021.
ROFL: Rolling on the Floor Laughing, ISBN-9781973300007, Amazon, 2017.
Seminole Rain: You don't choose destiny. It chooses you, ISBN-9798668502196, Amazon, 2020.
Septillionth: 1 in 10^{24}, ISBN-9798410762472, Amazon, 2022.
Software Development: Targeted Applications, ISBN-9798850653989, Amazon, 2023.
Software Recipes: Proven Tools, ISBN-9798815229556, Amazon, 2022.
Steam 2020: to 150 GPa and 6000 K, ISBN-9798634643830, Amazon, 2020.
Thermochemical Reactions: Numerical Solutions, ISBN-9781073417872, Amazon, 2019.
Thermodynamic and Transport Properties of Fluids, ISBN-9781092120845, Amazon, 2019.
Thermodynamic Cycles: Effective Modeling Strategies for Software Development, ISBN-9781070934372, Amazon, 2019.
Thermodynamics - Theory & Practice: The science of energy and power, ISBN-9781520339795, Amazon, 2016.
Version-Independent Programming: Code Development Guidelines for the Windows® Operating System, ISBN-9781520339146, Amazon, 2016.
The Waterfront Murders: As you sow, so shall you reap, ISBN-9798611314500, Amazon, 2020.
Weather Data: Where To Get It and How To Process It, ISBN-9798868037894, Amazon, 2023.